Proving Grounds

Proving Grounds

Project Plowshare and the Unrealized Dream of Nuclear Earthmoving

SCOTT KIRSCH

RUTGERS UNIVERSITY PRESS
New Brunswick, New Jersey, and London

Library of Congress Cataloging-in-Publication Data
Kirsch, Scott, 1967–
Proving grounds : Project Plowshare and the unrealized dream of nuclear earthmoving / Scott Kirsch.
p. cm.
Includes bibliographical references and index.
ISBN-13: 978-0-8135-3666-8 (hardcover : alk. paper)
1. Nuclear excavation. 2. Project Plowshare (U.S.) I. Title.
TA748.K57 2005
624.1′52—dc22 2005002579

A British Cataloging-in-Publication record for this book is available from the British Library.

Manufactured in the United States of America

For Bradley

Contents

Acknowledgments

This book tells the story of the rise and fall of an idea, nuclear earthmoving—the use of nuclear explosions for massive construction jobs like harbor and canal excavation—amid the laboratories and proving grounds of Cold War America. I could not have written it without the help of a great many people and institutions. These personal, intellectual, and professional debts have been gratefully accumulated in the course of a project that has led me across disciplines as well as across numerous time zones. It is a long list, but it is my pleasure to acknowledge them.

Proving Grounds began in the Department of Geography at the University of Colorado at Boulder, where it was stimulated by an atmosphere of critical debate and supported under a National Science Foundation grant (SBR-9628292). At crucial points in the research, I benefited from the insights and the arguments of Philip Deloria, John O'Loughlin, Brian Page, and Lynn Staeheli, whose readings of my work were all the more helpful because they rarely pulled their punches. I owe the most to Don Mitchell. Don invited my collaboration on his study of Project Chariot, a proposed nuclear earthmoving experiment in Alaska, which led to my interests in the larger story presented here, and he also shared archival data from his research on the Chariot environmental studies program. Always generous with his time as well as his criticism, Don provided ideas, encouragement, and friendship, which have all been invaluable not just in the development of this project but also in my own development as a scholar and teacher.

Positions in history of science programs at UCLA and the University of Oklahoma exposed me to a surprisingly different academic tradition, one from which I have continued to learn a great deal. My thanks to Ted Porter, for first enabling this move through UCLA's Center for the Cultural Studies of Science, Technology, and Medicine; and to Gregg Mitman, for sharing his interests in science and place with me through the Mellon Foundation program in Historical Intersections of the Biological and Social Sciences. Although I did not manage to write a word of this book during a year in the Science, Technology, and Society Program at Penn State University, I learned a great deal about teaching, and about the need to engage with natural scientists and engineers as well as with social scientists and historians, through my STS colleagues.

Moving across so many conceptual and geographical landscapes has been a fascinating but, inevitably, disorienting journey. I am fortunate to have made it with the kind support of colleagues who have helped out along the way and contributed to this work, whether in the form of comments on specific drafts, articles, or presentations; suggested readings; encouraging words; or good comradeship. Special thanks to Stuart Aitken, Bruce Braun, Michael Brown, Denis Cosgrove, Richard Devon, Lorraine Dowler, Gabriel Finkelstein, Colin Flint, Matt Hannah, Ari Kelman, Susan Millar, Katherine Pandora, Dick Peterec, Ralph Saunders, Anna Secor, Sharon Traweek, Jessica Wang, Peter Wissoker, Charles Withers, Lakshman Yapa, and the International Critical Geography Group. I am also grateful for the advice of editors and reviewers of the journals where scattered pieces of this volume have previously appeared in different forms: *Antipode*, *Ecumene*, *Osiris* (used by permission of the History of Science Society), and *The Professional Geographer*.

The Geography Department at the University of North Carolina at Chapel Hill has become a comfortable home for research, writing, and teaching, and an especially supportive one for junior faculty. A UNC Faculty Research and Study Leave provided critical time for renewing and rethinking my research. Also critical has been the support of colleagues and friends in Chapel Hill, and the stimulus provided by ongoing discussions over seminar tables, coffee, and beer. Special thanks to Altha Cravey, Arturo Escobar, Wil Gesler, Banu Gökariksel, David Havlick, Jon Lepofsky, Aaron Moody, Sally Nash, John Pickles, Carlos Tovares, Wendy Wolford, and Leo Zonn. Jun Liang, in addition to producing Figure 2.1, provided much needed technical assistance with many of the other maps and illustrations.

This book has also benefited from the work of librarians and archivists who have consistently made my life easier. I am grateful to staff at the Office of History and Historical Records, University of California Lawrence Livermore National Laboratory, in Livermore, California, and the U.S. Department of Energy's Nuclear Testing Archive (formerly the Coordination and Information Center) in Las Vegas, Nevada, the two collections on which I rely most heavily. Also contributing to this study were staff members at the collections of the Alaska and Polar Regions Department, Elmer E. Rasmuson Library, University of Alaska Fairbanks; the Australian Nuclear Science and Technology Organisation; and the Mandeville Special Collections Library, University of California, San Diego. Although the book is based mainly on these archival sources as well as news accounts and earlier scholarship, conversations with three scientists who were, in quite different ways, "present at the creation" of nuclear earthmoving provided a helpful triangulation of perspective and experience as I prepared the final manuscript for publication; my thanks to Barry Commoner, Milo Nordyke, and Herbert York.

It has been my good fortune to have twice met Audra Wolfe, my editor at Rutgers University Press, in Philadelphia. Thanks to Audra for her critical encouragement and myriad contributions to this book, and to the helpful editorial and production staff at the Press. I am also grateful to Mary Palevsky and anonymous readers who provided constructive critiques of an earlier draft.

It remains only to thank those closest to me. The unconditional support of my parents, Joy and Jeffrey Kirsch, has meant more to me than I have ever been able to express, except to provide the same for my own children. My daughter, Molly, shared a room with this book in its formative stages, and has grown up alongside it. My gratitude to Molly and her brother Hugh, however, is not for specific contributions to this project but for their success, at times, at keeping me away from it. Like no one else, my wife, Bradley, has suffered this work with me, and as a result it never really seemed like suffering. I dedicate the book to Bradley, for without her love, strength, and companionship so much less in my life would be possible.

Proving Grounds

If anyone wants a hole in the ground, nuclear explosives can make big holes.

—EDWARD TELLER, 1962

Introduction Geographical Engineering

In September 1964, the U.S. Congress authorized a five-year, 17.5-million-dollar study, to be conducted by the Army Corps of Engineers in collaboration with the Atomic Energy Commission (AEC), to explore the possibilities of blasting a new, sea-level canal through Central America, "whether by conventional or nuclear excavation."[1] According to preliminary estimates, the canal excavation could be accomplished by burying 300 to 310 nuclear explosives at varying depths for forty-six miles across the Darien region in southern Panama, and detonating them in a series of fourteen coordinated "row-charge" excavation cuts. Alternatively, to mention just one of several routes under consideration, 260 to 764 explosives, depending on the model employed, could be used to carve a channel across forty-three miles in northern Colombia at the southern tip of the American isthmus.[2] Excavating a new sea-level canal with nuclear explosives in either location would require about twenty-thousand times the explosive energy of the atomic bomb dropped on Hiroshima.

Looking back, it is difficult to say which part of the project demanded more hubris: the engineering problem of calculating the precise scale, layout, and depth of burial of explosives for the nuclear canal or the geographical problem of finding the right *place* for it, whether in Panama, Colombia, Mexico, or Nicaragua, on the basis of categories such as population density, proximity to cities or towns, vulnerability of nearby buildings to structural damage, and soil chemistry—in short, the expert construction of broad swaths of Central America as appropriate, and others not, for nuclear earthmoving. The map of projected fallout sectors for the nuclear canal project (figure I.1), later produced by the Lawrence Radiation Laboratory in Livermore, California, one of the nation's two nuclear weapons laboratories, extended the logic of this work by locating

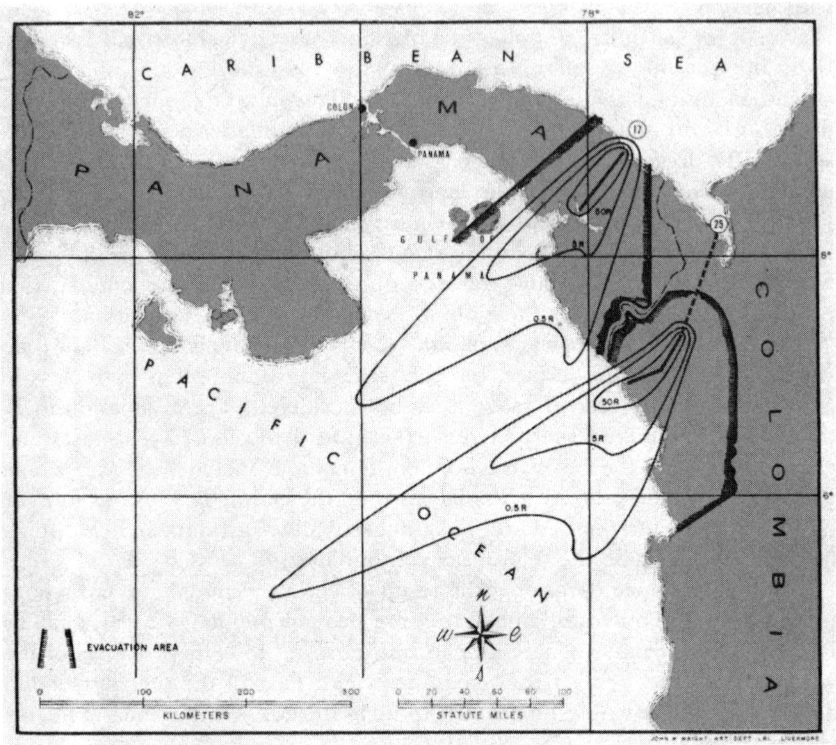

Figure I.1. Map of projected fallout patterns and evacuation areas (inside cross-hatching) for Route 17 (Darien Gap, Panama) and Route 25 (northern Colombia) of the proposed nuclear Central American canal. *Source: Edward Teller et al.*, The Constructive Uses of Nuclear Explosives *(New York: McGraw-Hill, 1968); Credit must be given to the University of California, Lawrence Livermore National Laboratory, and the Department of Energy (Atomic Energy Commission) under whose auspices the work was performed when this information or a reproduction of it is used.*

"evacuation areas" alongside favored nuclear canal routes 17 (Darien, Panama) and 25 (northern Colombia). Perhaps thirty thousand people, most of whom were living closely off the land, would have to be moved from the immediate downwind areas, according to estimates, at least until short-lived radionuclides decayed within operative permissible dose levels.[3] The map's greatest conceit, though, was the illusion of technical control that it evoked: the very idea that fallout sectors could be accurately drawn on a map before being produced in the landscape, and that evacuation areas encompassing *only* thirty thousand people for this earth-shattering event could be determined so precisely within much larger Panamanian, Colombian, and regional populations.

One effect of the authorization of the nuclear canal study in 1964 was to bolster the AEC's Project Plowshare, an experimental program launched at Livermore (now the Lawrence Livermore National Laboratory) in 1957 to investigate and promote the development of "peaceful" and "constructive" uses of nuclear explosives. Such applications were to include underground mining and natural gas stimulation, energy production (from explosions), and the production of heavy elements for use in scientific and medical research, but for most program supporters, *nuclear excavation* technology—earthmoving with hydrogen bombs for massive civil engineering projects—was initially the most promising, and indeed the most exciting, of the potential peaceful nuclear explosions. Among the most dedicated advocates of nuclear earthmoving was the famed physicist, defense commentator, and anticommunist Edward Teller. As Teller described it, "The ability of nuclear explosions to move vast quantities of earth and rock—and to move them cheaply—opens a new and important discipline: Geographical engineering. We will change the earth's surface to suit us."[4] In addition to the new Central American canal scheme, which had been under study at Livermore since 1959, the geographical engineering of continental river systems, dams, quarries, vast road cuts, and nuclear-blasted "instant harbors" on every continent but Europe would all become possible under Project Plowshare, or so it was argued.

Whatever the fount of Teller's optimism, it is obvious today that there was nothing inevitable about the success of geographical engineering. Since "nuclear cratering" bombs were detonated underground but, by design, broke the earth's surface in blasting away rock and soil, the explosions stirred up a great deal of radioactive dust and debris. Largely in response to these radioactivity hazards, and to the difficulties faced in their evaluation, the program in geographical engineering was challenged, almost from the start, on a variety of scientific, political, legal, and ethical grounds. In fact, despite their ongoing claims that "clean" (fallout-free) or at least cleaner bombs would be produced, and that nuclear excavation could be achieved without injuring the health of people and environments, Plowshare's program builders found that the distances between the map and the thing itself, and between AEC laboratories and the places-to-be-engineered outside them, were difficult paths to travel, open to contest, negotiation, and transformation at many points along the way. From the activist movements of an Inupiat village and dissident contract scientists in Alaska to the development of a new model for assessing radioactivity uptake in infants living downwind from Nevada nuclear testing to the ambiguous position of Plowshare in relation to international testing regimes, Livermore's experimental program in nuclear excavation navigated from controversy to controversy, continually transformed in the process. In 1970 the Army Corps of Engineers was forced, with regret, to drop what it called the "nuclear solution" from its plans for a new Central American canal, and the experimental program in nuclear excavation, though still incomplete, was defunded soon after. Redoubling its efforts in underground gas stimulation and shale oil recovery techniques, Plowshare persisted at reduced funding levels until the collapse of the AEC itself into its successor agencies in 1974. After that reorganization Plowshare withered into "paper studies" and was formally dissolved with the creation of the U.S. Department of Energy in 1977.

This book examines the work that went into putting geographical engineering "on the map" through Project Plowshare, and the relations of controversy and opposition that kept it, for the most part, out of the landscape. In part, it is a study of hubris and failure. It seeks to understand how the notion of "changing the earth's surface to suit us" with nuclear explosives made sense, for some, through specific discourses and ways of knowing, and through the cultivation of an "engineering standpoint," as Teller would later describe it, from which politics and emotional issues were to be excluded from applied scientific considerations.[5] *Proving Grounds* tells the story of how the new science and technology of nuclear earthmoving were constituted, alongside this engineering standpoint, in the collective work of many scientists, bureaucrats, politicians, engineers, journalists, activists, educators, and others, amid the institutional context of what was, in reality, a highly politicized Cold War program. But by investigating the work of these actors and tracing the relations among them, *Proving Grounds* also seeks to better understand the obstacles that nuclear excavation ran up against. Resistance to Plowshare projects, like challenges to the larger American nuclear testing program, generally resulted from more precautionary—and at times, explicitly oppositional—scientific and political projects, including the contributions of ecological and public health scientists working both within and outside AEC channels.

Thirty-five years after the American nuclear earthmoving program was effectively shut down with the termination of the nuclear canal study, the idea of a weapons laboratory engaging in grandiose Atomic Age civil engineering schemes may appear quaint. But the important questions about exactly how the program was undermined, and for that matter, how it persisted for as long as it did, and to what effects, are not approachable without close historical study.[6] A geographical history of the controversial program, which raises questions about how the endeavor was shaped by variation and contest from science to science, institution to institution, and place to place, can tell us a great deal about the politics of science in environmental transformations and about the dynamics of a Cold War technological failure (and, for a time, laboratory success) that was both spectacular and banal. Before this story can be told, however, there are a few questions to address: why a *geographical* history of an experimental program, and why *this* experimental program? What can we learn from the apparently strange case of nuclear earthmoving that, from today's perspective, may appear as merely a bizarre misstep in the modern narrative of cumulative scientific-technical progress?

The Strange Case of Project Plowshare? Dynamics of Controversy and Failure

The temptation is to dismiss the grand possibilities envisaged for nuclear earthmoving at the outset. One might quickly surmise that the program must have failed because nuclear cratering explosions posed severe threats to human and environmental health or, perhaps, as Plowshare proponents often rationalized challenges to the program, because it became "politically unacceptable." Neither of these perspectives on the past is necessarily false, but as historical explanations, they are akin to what the anthropologist and philosopher of science Bruno Latour would call *ready-made:* they use the settlement of a controversy to explain how and why the controversy was settled.[7] In effect, the outcomes of complex disputes, such as contests among scientists over the dangers of low-level radiation from fallout, are used to explain the processes that produced those very outcomes. This retrofitting of scientific and technical knowledge into contexts in which it was not yet available, or more complexly, into contexts in which the facts were agreed upon by some, but contested by others, results in an impoverished sense of agency for all manner of social and political relations, including those of scientific work. Nuclear earthmoving failed, in this sense, because it was propelled from the laboratory into the world as, inherently, a failure.

I begin instead with the premise that the failures of geographical engineering, and of Project Plowshare more broadly, were not predetermined (whatever the eventual health and ecological outcomes of the projects may have been), and were hardly predictable for Plowshare scientists and supporters, many of whom were entrenched in leadership positions in the AEC nuclear science, weapons, and technology complex during the late 1950s and throughout the 1960s. The fact that a more active Soviet peaceful nuclear explosives program persisted until 1989, amid quite different political, scientific, and medical contexts and norms, underscores this point, even as the Soviets never accomplished their own geographical engineering dreams of irrigating Central Asian deserts with Siberian waters. By examining not only how Livermore physicists and AEC radiation experts may have managed to convince *themselves* that certain excavation projects could be carried out safely but also how these actors responded to the problems posed by issues of scientific uncertainty, risk, and the dynamic nature of contexts and norms, we can explore questions about the program that notions of ready-made failure tend to preclude. How did the twists and turns of the program relate to controversies over the health effects of nuclear weapons testing more generally, and to related debates during the late 1950s and early 1960s over testing moratoria, treaties, and limits? How did the "technically feasible," for Plowshare scientists and engineers, ultimately become the politically unacceptable?

One claim to be elaborated in this book is that the controversial history of nuclear earthmoving was made to turn, in critical ways, on both ecological and spatial issues, including spatial practices of territoriality, boundary making, and geographical mobility. For while nuclear earthmoving may have been an utter failure for the AEC, it did not fail everywhere at once. The program was pursued for almost fifteen years, that is, but only in certain places and in certain communities where the ideas, funding, and authority of the program held sway. More than a story of failure, it is also one of enduring resilience in the face of multiple challenges. Was success or progress achieved in the nuclear earthmoving program on different registers, whether technical or political? Approaching these issues requires a survey of the varied work that went into the program: although Plowshare, as will see in the next chapter, was constituted around a number of political, entrepreneurial, and perhaps military aims, it nevertheless remained the goal of most project scientists and engineers for the duration of the program to put geographical engineering into practice. We will examine the work of the Plowshare scientists and supporters—at the height of the Cold War, amid widespread public anxieties over nuclear-test fallout, a nuclear arms race, and the possibility of nuclear war—as they engaged in what was at once a scientific and technological, political, and moral mission: the effort to turn the bomb into something good.

The AEC, garnering appropriations through the congressional Joint Committee on Atomic Energy, spent at least $770 million (in constant 1996 dollars) on Project Plowshare between 1957 and 1974. As a percentage of the total costs of U.S. nuclear weapons programs, now measured in the thousands of billions, this sum is not great, but it was substantial enough investment to support an ambitious research and development program. From 1959 to 1973, some 125 to 225 full-time-equivalent employees per year labored on Plowshare at Livermore. In the earthmoving field, this work included experiments with nuclear bombs and chemical high explosives; efforts to develop precise emplacement techniques, developing mathematical "cratering curves" for predicting crater size, fallout distribution, and other effects; efforts (often inseparable from related military projects) aimed at producing cleaner nuclear bombs for excavation; excavation experiments involving extensive site selection procedures, promotional efforts, and collaborations with government, industrial, and professional partners; and the development of technical feasibility and "safety feasibility" studies for model projects and nuclear cratering experiments. Much of this work was designed conceptually at Livermore, where technical direction of Plowshare was located among senior laboratory scientists, and where an Army Corps of Engineers "Nuclear Cratering Group" established a home base for collaboration on the Central American canal study and other projects. But the work was also spread across many other institutions within the classified community of AEC advisory committees, testing and radiation safety operations, and contract engineering firms. Under the Plowshare program, the AEC also supported a number of trailblazing bioenvironmental and radioecological studies, initiated as contract fieldwork intended to gauge how radioactivity from nuclear earthmoving explosions would move through ecosystems.[8]

Before planetary engineering jobs like the Central American canal could be undertaken, however, even the most optimistic Livermore scientists believed that an extensive experimental program would be necessary to develop nuclear excavation techniques properly, under putatively controlled conditions. Carrying out such a program would also produce its own set of needs, not least the demand for geographies to engineer, or at least to experiment with. Some of the nuclear earthmoving experiments could be conducted at the AEC's test site in Nevada, but Plowshare scientists insisted that they must also test nuclear excavation in the "pragmatic engineering world in which the problems of the environment must be met and solutions found."[9] Livermore's researchers thus endeavored to take their nuclear-blasting concept to places ranging from northwestern Alaska to California's Mojave Desert to Western Australia for "demonstration projects" that were designed both as experiments producing useful scientific-technical data and as spectacular public demonstrations of the efficacy and safety of nuclear earthmoving. Yet while six different nuclear cratering tests were carried out at the Nevada Test Site between 1962 and 1968, Plowshare's program builders never managed to conduct their nuclear earthmoving experiments or model projects anywhere *else*.[10]

The failure of the Plowshare scientists to make what might be seen as a fundamentally geographical move, that is, to take their capability of blasting big holes in the ground from one experimental space to another, raises further questions. How was it that what could be called safe or true in one place was not so in another? How did disciplinary and institutional differences among the sciences intersect with contests over the authority of AEC scientists to predict the environmental and health consequences of cratering explosions?

How did socially produced spaces and boundaries, and circulations across boundaries, serve as practical solutions to both the political and the epistemological problems of nuclear earthmoving?[11] And what does the unrealized dream of geographical engineering tell us about the place of science and scientists in environmental politics at a moment in U.S. history—traversing (and intersecting with) Rachel Carson's 1962 *Silent Spring* and the 1969 National Environmental Policy Act—when environmental politics and values were profoundly reconfigured? In the laboratories and proving grounds of Project Plowshare, the histories of experiment and environment meet.

The value of these questions does not rest on whether or not Plowshare—a project that is interesting in part because it was on the margins of possibility—can be made to stand for science itself. In the context of the American military-industrial-scientific complex during the Cold War, however, to insist on clean distinctions between basic and applied, science and technology, or military and commercial, would be not only an empirically difficult task but an ideologically misguided one as well.[12] Rather than a case in which a small percentage of "pure" research stand for the whole of science, Plowshare offers a chance to examine precisely the intermingling of scientific, technical, political, security, and cultural discourses and logics. In Plowshare we can see how these logics, and the institutions in which they were rooted, have at times been effectively challenged, and in some ways transformed, on scientific, ethical, and political grounds.

Describing the "strange case of *la mer intérieure*," an unrealized French colonial plan for producing an inland sea in the Sahara, geographer Michael Heffernan not only tells the story of an abiding faith in scientific and technological progress among nineteenth century French colonials that sustained this scheme for many years, but also explores the ways in which the very idea of progress was "transformed in the process of its own development."[13] Plowshare also embodied a certain vision of progress, most spectacularly evident in the nuclear earthmoving concept, and it too is a story of the transformation of this dream, though perhaps not in the minds of its protagonists. Rather, in the limits imposed on Plowshare we glimpse a moment in which many scientists, citizens, and activists began to take seriously the consequences of environmental transformations *before* they took shape, and thus an altogether different idea of progress emerges.

A Geographical History

"A good nuclear engineer," Edward Teller once explained to educators, "has to study lots more than the nuclear physics. He has to study the modern miracles of the computing machines and their uses; he has to study hydrodynamics and geology; in fact, he has to study geography because how else could he engage in geographic engineering, the enterprise intended to change the geography so that it should better serve human purposes?"[14] But rather than providing simply another set of useful information for the nuclear engineer, we will see that geographical knowledge—that is, knowledge of spatial distributions of flora, fauna, geology, agricultural and hunting practices, and knowledge of the relations between these spatial distributions and ongoing processes of ecological and historical change—would itself become disruptive to the Plowshare nuclear earthmoving program, complicating the narratives of scientific and technological progress with which Plowshare had originally been coded and sold.

Whereas Teller and his colleagues tried to reduce geography to an object of engineering, *Proving Grounds* attempts a geographical history of their work predicated on more nuanced, relational understandings of place, space, and the representational practices of science and cartography through which spaces have been reconfigured. In the next chapter, I locate Plowshare's origins within the institutional context of the American Cold War military-industrial-scientific complex and under the conditions of controversy and uncertainty from which the excavation program was pursued. The chapter examines the institutional development of two of Plowshare's key laboratory spaces, the Livermore laboratory and the Nevada Test Site, in order to situate Plowshare's emergence in the late 1950s as both a research and development program and a kind of public information strategy.

Chapter 2 traces the initial planning of the nuclear excavation program at Livermore. Early project goals blended the aims of Plowshare as science and spectacle as the program moved to Alaska to enact Project Chariot, the nuclear harbor scheme that was to be the program's first demonstration project. Challenges to this plan emerged that were undertaken initially by biologists from the University of Alaska, Fairbanks. Chapter 3 then explores the politically charged geographies of authority and controversy that characterized Chariot scientific work and political activism alike, examining the development of oppositional movements around two overlapping strains of resistance: that of dissenting biological and geographical scientists enlisted in the AEC's Chariot environmental studies program and that of Eskimos from Point Hope, Alaska, a village located some thirty-one miles from the proposed ground zero. The Chariot controversy is resolved, as the chapter concludes, by a geographical retreat to the Nevada Test Site for the first major Plowshare excavation experiment, Project Sedan.

The Limited Test Ban Treaty (LTBT), formally the Treaty Banning Nuclear Weapon Tests in the Atmosphere, in Outer Space and Under Water, was signed by the United Kingdom, the United States, and the Soviet Union in

1963. In addition to prohibiting atmospheric and underwater nuclear testing (while allowing for underground testing), the treaty banned the movement of radioactive debris across international borders, a matter of particular concern for advocates of nuclear earthmoving. Nevertheless, after the ratification of the LTBT, and even as new controversies about the dangers of ingested radioiodine from nuclear fallout erupted in both scientific journals and congressional hearings, Plowshare received unprecedented funding increases for its nuclear excavation program, largely in connection with the five-year feasibility study of the nuclear Trans-Isthmian Canal project. Chapter 4 focuses on how the experimental program in nuclear excavation was carried out, at least to an extent, within the restricted but functional spaces of the Nevada Test Site. It also examines how "outside" controversies nevertheless structured the nature of the knowledge produced from nuclear cratering tests in Nevada. Despite growing evidence to the contrary, Plowshare scientists remained confident that the environmental problems associated with fallout and geographical engineering could be technically and pragmatically solved. In addition to the canal study, they pursued several smaller demonstration projects around the world throughout the 1960s. Chapter 5 builds on issues raised in the previous chapter to investigate how these endeavors were pursued by Plowshare and its partners through discourses of engineering feasibility, on the one hand, and of "progress-through-experimentation," on the other, in which the sheer generation of data could be interpreted as an experimental success.

Despite some success convincing university physics and engineering departments, professional engineering societies, and potential U.S. and international partner agencies that geographical engineering was a viable enterprise, in December 1970, after fourteen years of research and promotion, the experimental program in nuclear earthmoving finally lost its currency among government patrons. The concluding chapter traces the final moments of a demise many years in the making before asking what should be written on Plowshare's tombstone. The chapter reflects on issues in the politics of science, technology, and environmental health raised by the Plowshare program in order to forge practical lessons, from this quite recent history, to inform our future politics of knowledge and environmental transformation.

Chapter 1 Origins of a Cold War Experimental Program

ONE STORY EXPLAINING the origins of Project Plowshare begins with the Suez crisis in the fall of 1956. As Edward Teller and three colleagues later described it for readers of their university textbook, *Constructive Uses of Nuclear Explosives*, it was following the Egyptian blockade of the Suez Canal that a small group of weapons scientists convened at the Atomic Energy Commission's Livermore laboratory. The meeting "raised the possibility of cutting another canal through friendly territory with nuclear explosives."[1] One coauthor, former Livermore test director Gerald Johnson, later remembered the meeting: "we simply pulled out a map, developed a profile from the Gulf of Aquaba to the Mediterranean and rapidly decided in principle it was technically possible."[2] And yet as Teller and Johnson both knew by the time they wrote their textbook, there was a world of difference between what was "technically possible" from a conference room in Livermore and what was actually possible in the landscape itself; a profile, quite obviously, is not a canal. Still, although this nuclear solution to the Suez crisis was never seriously considered at the laboratory, let alone in Washington, it did make a compelling story for public advocates eager to stress the utility of "peaceful nuclear explosions" and for those, like Teller, wishing to inspire a certain kind of technological and geographical imagination.[3]

Whereas the textbook account locates the impetus for Plowshare outside the laboratory, with the scientists at Livermore reacting to serve the national and global needs of a Cold War crisis, this chapter focuses instead on the scale of the laboratory itself. How was Plowshare created as an experimental research program in 1957, and why? How did the Livermore scientists, among others, act as authors, program builders, and activists for Plowshare under conditions of political and scientific uncertainty related to their work?

These conditions of uncertainty were evident in the emergence of highly public, interrelated debates over the harmful effects of radioactive fallout from nuclear testing, on the one hand, and the merits of international nuclear test ban treaties and testing moratoria, on the other. When Teller, Johnson, and their Livermore colleagues contemplated putting nuclear explosives, the principal product of their laboratory, to work in blasting an alternative to the Suez Canal, the AEC and its weapons laboratories had already presided over more than eighty test explosions, most of them set off in the atmosphere or underwater, at proving grounds in the Pacific Marshall Islands and the Nevada desert. The pace of this testing, and of the scientific research and industrial production processes to which it contributed, had accelerated dramatically following the first Soviet atomic bomb test in 1949, signaling the emergence of a high-tech arms race apparently predicated on the results of such experiments. And yet by February 1957, when a small symposium was convened at Livermore to explore the possibility of a formal program for developing nonmilitary uses of nuclear explosions, resistance to nuclear testing in general was becoming more widespread. Opposition, which included legal action by Nevada ranchers seeking restitution for damaged sheep, scientists organizing antitesting petitions, congressional hearings and serious international considerations of testing bans and limits, threatened to undermine the practice of nuclear testing as the legitimate and necessary source of knowledge about nuclear weapons, and thus perhaps to cut off the principal lifeblood of the AEC weapons laboratories at Los Alamos and Livermore as well.[4]

This chapter traces the origins of Project Plowshare, and the nuclear excavation program in particular, by looking to those who created the program, the scientific, institutional, and geographical spaces in which they worked, and the controversies and conflicts "outside" those spaces that constrained—and in some ways enabled—that work. We turn next to the 1957 Livermore non-military uses symposium, and then to the political, scientific, and institutional settings that made the meeting possible.

Livermore Symposium

Whatever the role of the Suez crisis in the germination of Project Plowshare, Teller was right to remember Livermore as the site of the idea of the peaceful nuclear explosion or, rather, as the place where the idea was transformed into a working experimental program.[5] How this transformation should be accomplished, and who would benefit from it, were among the chief problems taken up in a joint "AEC Weapons Laboratory Symposium on Non-Military Uses of Nuclear and Thermonuclear Explosions" held at Livermore from February 6 to 8, 1957. With personnel drawn from Livermore, Los Alamos, the Sandia Corporation (a New Mexico engineering laboratory and AEC contractor), and the AEC's San Francisco Operations Office and its Division of Military Applications (DMA), which presided over a good deal of the weapons laboratories' annual budgets, the symposium provided a social space—for those with AEC security clearances and access to its classified data—for deliberative discussion and critical feedback. Here, among insiders, was Plowshare's first proving ground.

One of the meeting's key outcomes was a summary, written by Livermore physicist Harold Brown for laboratory director Herbert York and distributed to more than two dozen personnel spread across the participating institutions, which condensed discussion and conclusions from the symposium, technical and political, into a five-page outline. The report described the prospects for the program as "exciting," "useful," and "promising." Even so, concerns over fallout risks were signaled; Brown advised, in the first point of his memo, that "an effort even more exhaustive than now under way to reduce the fission activity connected with thermonuclear yields, both large and small, is very strongly indicated. This point came up continually during the meetings, and was universally subscribed to."[6]

Thermonuclear (fusion) explosions, or hydrogen bombs, a proven part of the AEC arsenal after 1952, had to be triggered by a fission explosion within the same device, from which highly toxic, long-living radionuclides such as strontium-90 and cesium-137 were known by-products (the original atomic bombs tested in New Mexico and exploded in Hiroshima and Nagasaki were entirely fission bombs). Because hydrogen bombs were potentially thousands of times more powerful than fission explosions, by reducing the fission component (or fission-to-fusion ratio) of thermonuclear explosives, scientists hoped to substantially reduce fallout even in much larger explosions.

With this caveat of reduced fallout in mind, symposium participants freely speculated about possible uses for nuclear explosives in mining operations, for stimulating oil flow, and for scientific experiments, perhaps in connection with the upcoming International Geophysical Year, through which underground and high-altitude explosions could be used for generating geophysical data. But participants saw the most imminent potential in the area of nuclear excavation, which was "extensively discussed." As Brown put it, "it appears that there are few technical obstacles to this use if the level of cleanness discussed above can be reached." And yet, although it seemed clear that the costs of nuclear explosives would far and away undercut those for conventional high explosives for such massive scale earthmoving jobs, it was also clear, at least for the time being, that there would be significant limits on *where* nuclear excavation could be carried out: "With present ratios of fission to total yield, however, the fallout and related political problems appear to most of us insuperable for any project in or near an inhabited area. Fallout from a substantial yield, in the hundreds of KT [kiloton] range, which vents the surface, as excavatory shots which dig a hole to the surface which is open to the surface must do, will extend for a distance on the order of 100 miles. Until the *very* low fission yield devices are available, then, one is confined to uninhabited areas."[7] Just how such remoteness from population would be defined, and exactly how low a fission yield made for "safe" nuclear excavation, were questions on which the history of the experimental program in nuclear excavation would be made to turn.

The considerations of the gathered scientists and engineers were not limited to technical matters. It was also clear to those attending the symposium that the success of such a program, like that of nuclear testing more generally, depended on some measure of public assent. Yet here too, according to Brown, the group was optimistic:

> It is plain that many of the difficulties involved in the non-military uses are not technical but have to do with questions of public opinion or with such nuisances as broken windows. The governments faces such nuisance problems in other activities also, but the situation is greatly complicated by the public fear of nuclear and thermonuclear bombs (quite justified) and of tests (less justified). The suggestion that such explosions may have non-military uses may not encounter as much of the public hostility produced by fear, and in fact such programs if successful might produce a change in attitude.[8]

Whatever the merits of the suggestion—that people whose windows were shattered by nuclear air blasts would be less hostile to nuclear tests if they were "peaceful" air blasts, and whether or not the heightened public fears of atmospheric nuclear weapons tests were justified in 1957, it is clear that public relations goals (that is, the politics of changing attitudes) were woven into the fabric of the program from the start.

But the most important question to come out of the symposium, as Brown saw it, was "what the nature and magnitude of subsequent effort in the area of non-military uses should be"—how the ideas hashed out at the meeting could be integrated into a concerted, institutionalized peaceful nuclear explosives program. Some data on "nuclear cratering" could be produced from "mocked-up tests" with chemical high explosives, while "paper studies," Brown admitted, would "of course be the cheapest of all, but they tend[ed] not to lead anywhere so long as relevant experimental data are missing." Nuclear tests were expensive though, and the vigor with which the program would be pursued thus ultimately depended on the matter of financial support. Brown stressed, however, that the program would not drain funds from weapons development programs, the laboratory's primary mission. Rather, by adding to the fund of concepts, data, and methods in nuclear explosives development, especially in the goal of producing cleaner nuclear explosives (considered desirable by the military as well as by advocates of unabated nuclear testing), the nonmilitary uses program "if suitably funded, would undoubtedly act to increase rather than decrease our weapons design activities." The competition for resources, in other words, was not viewed as a zero-sum game. Brown proposed the establishment of a formal program for nonmilitary uses, on the one hand, and the expansion of support for "such related projects" as clean weapons design and deep underground nuclear tests in Nevada, on the other.[9]

From the beginning, then, the aims of the would-be program were defined by two important boundaries. First, at least until cleaner bombs could be made, there would be geographical limits to where the scientists could test their nuclear earthmoving concepts. Second, research and development efforts for nonmilitary and military explosive designs—though intimately related in scientific theory and experimental practice—would be set apart in public information policy. A more detailed public information program would accompany the peaceful nuclear explosions research than was typical of the weapons programs, without any additional disclosure of explosive design secrets. Working through the DMA and with the support of prominent senior scientists such as Harold Brown, Gerald Johnson, Teller, and Berkeley physicist Ernest Lawrence, the newly named Project Plowshare was formally approved by the AEC in June 1957; the program was to be "technically directed" from Livermore and administered through the AEC's San Francisco Operations Office.[10] Although Teller would quickly emerge as a public spokesperson, it was Brown who was most active in establishing Plowshare, and Johnson who was charged with leading the Livermore program in its formative years.[11] But why was the program put forward at this time and in this place? Who expected to benefit from it? In what kind of social and spatial settings was it conceived?[12]

Spaces and Boundaries of AEC Weapons Science

Plowshare retained its distinction as a Livermore program as it emerged into a wider network of AEC actors, resources, and classified sites, places characterized by particular rules of secrecy and security, disciplinary conventions, and institutional cultures. As part of the American military-industrial complex—a phrase made famous by President Dwight Eisenhower in his outgoing warning to the American public—the AEC (and by extension, Plowshare) was part of a new organization of productive relations among government, military, scientific, and capitalist industrial actors collaborating on the work of Cold War arms production. The mass mobilization of industry for technological weapons production in general, and for atomic and nuclear explosives in particular, was a process which, to paraphrase Marx, sweated money from every pore. Scientific research was among the military-industrial complex's greatest beneficiaries. As historians and other scholars have observed, the new economy of Cold War science had a particularly strong geographical expression in the western states, where federal funding—bolstered by laboratory budgets for Livermore and Los Alamos as well as Lawrence's Radiation Laboratory in Berkeley (all managed by the University of California) and the Jet Propulsion Laboratory in Pasadena, California—reached an astonishing 1,600 times pre–World War II levels.[13] But there is more to this geography or spatiality of knowledge production than the matter of regional location. AEC weapons science was itself shaped by a short but intensive history of boundary making in the governance of its scientists, their secrets, and their workplaces. To explore the boundary making and policing that helped to shape the AEC in turn into an organization with clearly marked "insides" and "outsides" we must briefly turn to the storied site of its origins: the Manhattan Project.[14]

It was during the winter of 1939–40, one year after physicists learned how to split the uranium atom, that the Roosevelt administration, at least partly at the behest of European émigré scientists who were concerned over the possibility of a German atomic weapon, began to put together a research program to explore the military uses of atomic energy. In doing so, the administration brought to bear the resources of the American state on the science of atomic physics, and at once, it imposed a new boundary of secrecy on that field. As Peter Bacon Hales argues in his history of the Manhattan Project's "atomic spaces" at Hanford, Oak Ridge, and Los Alamos, this code of secrecy, constituted under fledgling federal agencies including the Uranium Committee, the National Defense Research Committee, the Office of Scientific Research and Development, and ultimately the Army Corps of Engineers' Manhattan Engineer District (a.k.a. the Manhattan Project), was a major departure from the way physics had normally been practiced as an explicitly public scientific discourse.[15] The splitting of the atom, after all, had been explained publicly in January 1939, and unlike, say, chemistry, with its already established links to private industry and commerce, knowledge production in physics had always depended, at least to an extent, on open and public scientific debate. But the introduction of military strategies of compartmentalization had repercussions both in the nature of the scientific work carried out in the atomic sciences and in the geographies created to facilitate that work.

"Compartmentalization" was the master strategy of General Leslie Groves, director of the Manhattan Project, for maintaining secrecy and control of a colossal project, employing up to 125,000 people between 1943 and 1945, primarily in construction, engineering, and production operations for the uranium enrichment facility at Oak Ridge, Tennessee, and the plutonium production plant at Hanford, Washington, while only a handful of project managers, let alone workers, knew what it was they were actually building. For example, most of those working on any aspect of the project were allowed to speak only to their superiors about their work, not to each other. As Hales shows for areas as diverse as labor relations, medicine and worker health, and pollution ecology, compartmentalization was a means of consolidating knowledge and power upward in a strict hierarchy of authority and control.[16] But for the scientists working on the project from some thirty different university laboratories and working groups across the United States, the imposition of new procedures associated with compartmentalization—whereby communications between laboratories, and at times even between programs at the same laboratory, essentially had to be channeled through the hybrid military-government-scientific agencies overseeing the project—was a major hindrance to the efficiency of the program. And so the idea for a central scientific laboratory on the remote New Mexican mesa at Los Alamos, where top physicists could work out many of the remaining theoretical problems associated with the development of the fission weapon itself, emerged in 1942–43 at the urging of Berkeley physicist J. Robert Oppenheimer and others as something of an afterthought for the Manhattan Project. As Hales describes, Los Alamos was a "geographical solution" to the problem of compartmentalization by creating "an ersatz university environment, an ivory tower within which those scientists best able to push the project along could work closely, sharing a common purpose, isolated from distraction. With security systems externalized, in the form of perimeter guards and protected gates, more complete background checks, and rigorous exclusion of all but the most unimpeachable, compartmentalization could then be relaxed, and the scientists could return to their arguments and open transmission of ideas."[17] While information communicated between Los Alamos and other research centers in Chicago, Berkeley, or Oak Ridge still fell under procedures of compartmentalization, the "externalization of security"—the deliberate production of insides and outsides at the Los Alamos laboratory—was a successful strategy of spatial concentration that countered some of the effects of the compartmentalization of information. But it was not clear who would control the geographical and intellectual spaces of American nuclear weapons science after the war.

During the Manhattan Project, atomic physicists had already chafed against the practice of having their work subject to military authority and secrecy rules, and the use of the bomb against the Japanese had confirmed the gravest misgivings of many project scientists. Widespread conservative (and military) support in Washington for the May-Johnson bill in the fall of 1945—a proposal for maintaining military control of the atomic program along with the codification of stringent new security and secrecy laws for program scientists—provided a focal point of opposition for the emerging "atomic scientists' movement." This movement, composed largely of younger Manhattan Project scientists impelled into activism, in no small part, as an ethical response to their own participation in the endeavor, successfully rallied opposition to May-Johnson among the public and in Congress and sought support instead for the international control of atomic energy and all fissionable materials. Concerned over military patronage of science in general, and of the atomic program in particular, the scientists' movement advanced what historian Jessica Wang has called a "progressive left politics of science" during the immediate postwar years, characterized by a broad commitment to internationalism, especially international control of atomic energy, and to a sense that publicly funded scientific research must be responsive to social needs.[18] There were no real atomic "secrets" as such, the scientists' movement argued, that would not eventually be learned by other countries through science (rather than espionage), and so a new international atomic order was the best hope for preventing both an international arms race and a devastating nuclear war.

Enlisting public support through speeches and the popular media (and later through the movement's *Bulletin of the Atomic Scientists*) and lobbying Senator Brien McMahon and others on the Senate Special Committee on Atomic Energy, activists in the scientists' movement not only helped to scuttle May-Johnson but also consulted extensively on its would-be replacement, the McMahon bill, in the autumn of 1945. Scaling back the extreme security measures evoked in May-Johnson, the McMahon bill proposed a civilian Atomic Energy Commission to manage the atomic energy program on a national scale while still leaving room for future international agreements, even stating that the AEC would not engage in weapons-related research if it violated any future international arrangements.

Yet by the time the McMahon bill made it through Congress as the Atomic Energy Act of 1946, it was a quite different piece of legislation, arguably a compromise between the poles of civilian and military control. The Atomic Energy Act did create a civilian Atomic Energy Commission, to be run by five commissioners appointed by the president, to inherit the Manhattan Project's atomic spaces, along with the information, skills, and infrastructure accumulated in them. The commissioners would be informed by a General Advisory

Committee (GAC), composed of nine scientists and engineers, also to be appointed by the president, which would analyze AEC policies and make recommendations, while the AEC, in turn, was to report justify its appropriations to a congressional Joint Committee on Atomic Energy (JCAE).[19] Far from the social and internationalist agenda pursued by the scientists' movement, the primary purpose of the new AEC was to be weapons production, much as the Department of Defense had wished, and McMahon had been amended to include many of the same unprecedented secrecy, loyalty, and classification provisions that the scientists' movement had objected to in May-Johnson. In addition, the legislation was devoid of any provisions for substantial public health and safety monitoring or regulation from outside the AEC. What emerged in practice was a new hybrid of state-military-scientific authority—structured by expertise in a narrowly defined scientific-technical sphere, and by privileged access to information—so that even the movement's apparent victory of civilian control was questionable. Three of the commission's first five "civilian" general managers of the AEC stepped down from active military duty to take the position, while 70 percent of expenditures during the commission's first fifteen years were administered through the DMA, which was directed by active military personnel.[20]

Scientific and military interests learned to accommodate each other within the AEC, or at least those scientists that remained there accommodated their work to a culture of military patronage. Some Manhattan Project scientists with qualms about weapons work returned to their universities after the war; others, as Wang puts it, forged "an alliance with the emerging national security state, and, despite moments of profound ambivalence, . . . perpetuated the continuing mobilization of the physical sciences for military purposes during the early Cold War years." At the same time, she argues, the rise of anticommunism during the late 1940s and early 1950s, and with it the hounding of atomic scientists working at the national laboratories through the loyalty-security system, put liberal and progressive scientists on the defensive; whatever the actual possibilities were for the atomic scientists' movement's early internationalist aims, any political opportunity was effectively closed out during the first years of the Cold War.[21] By the time Oppenheimer's security clearance was infamously stripped in 1954, incoming generations of weapons scientists were already being socialized into the weapons laboratories with the cause of nationally based weapons development, and the conventions of classified science, firmly established a priori, for their chief scientific mission and for their livelihoods. The most important boundary to emerge after Hiroshima and Nagasaki was not between the scientists and the generals, but rather between the AEC and the world outside.

Growth

Epochal notions of an Atomic Age, in which people everywhere suddenly faced the new threat of world annihilation, are not without a certain verity, but they also obscure a much more specific geography of uranium mining and weapons production, research, and testing, nodes of production in an actually existing atomic age. As the Manhattan Project's "secret cities" in Washington, Tennessee, and New Mexico entered public knowledge following the bombing of Hiroshima, it quickly became evident that the scientific and industrial work of making the atomic age, with all of its global implications, was a decidedly local and regional phenomenon as well. The AEC inherited its infrastructure from the Manhattan Project in the form of plutonium and uranium fuel production facilities at Hanford and Oak Ridge, and a new space for testing among the atolls of the Marshall Islands, in roughly the middle of the Pacific Ocean, where the first postwar atomic tests were held under Manhattan Project authority at Bikini Atoll in 1946. In addition to maintaining the Los Alamos site as a laboratory devoted chiefly to weapons research, to be managed still by the University of California, the AEC would also support new "national laboratories," maintaining active research programs in areas of high-energy and solid-state physics, reactor development, and biomedicine, at sites including the University of Chicago's Argonne laboratory (the suburban successor to the Manhattan Project's Chicago metallurgical lab), the Clinton Laboratories at Oak Ridge, the Brookhaven Laboratory on Long Island, managed by a consortium of fourteen eastern universities, and Lawrence's University of California Radiation Laboratory at Berkeley.[22]

As the Cold War flared during the late 1940s and early 1950s with the coming of the Soviet atomic bomb, China's communist revolution, and U.S. involvement in the Korean War, the AEC quickly moved to expand its spaces of nuclear weapons production and research, building more laboratories, more factories, and more spaces for experimentation and the development of new capabilities.

NEVADA TEST SITE

While the nuclear complex was built in secret during World War II, the military appropriation of vast tracts of land and air space in the state of Nevada for pilot training and gunnery ranges proceeded publicly and without controversy. From its pattern of land tenure, with roughly 80 percent of the state constituted by federal lands, to its relatively sparse population to its dry climate and three hundred clear flying days per year, Nevada was, from the army standpoint, naturally suited to use as a gunnery range. Its desert basin and range lands were also well situated geographically as a staging area for war in the Pacific, located close to the coast but nested behind the Sierra range. Before the war had begun, President Roosevelt had signed an executive order, under authority of the Army Appropriations Act, to create the Las Vegas Bombing and Gunnery Range out of an expanse of three and a half million acres, located sixty-five miles northwest of Las Vegas, on an area still claimed by Native Americans from the Western Shoshone Tribe.[23] Although the range, like many wartime military facilities in the United States, had been decommissioned by 1947, it was reactivated just two years later under the aegis of the newly formed air force. Soon a new continental nuclear proving grounds would be grafted onto this landscape as well.

As early as 1948, after experience with the extended supply lines needed for Pacific tests, the Armed Forces Special Weapons Project undertook a study of possible atomic test sites to be located in the continental United States.[24] This study was put aside in early 1949 "for possible later use in the event of an emergency."[25] In July of 1950, the month after U.S. involvement began in Korea, the AEC revived the survey, arguing that "a national emergency is, at least, possible."[26] A continental test site, to be used for "low-yield" blasts while the larger bombs would still be detonated at the Enewetak Proving Ground in the Marshall Islands, would be "capable of serving economy in terms of time, personnel, equipment and money, and, above all, [would] expedite the weapons development program."[27] With these priorities in mind, the work of site selection was reduced to four central criteria: radiological safety (including population density and meteorological conditions), "physical requirements" (size, climate, topography, location, and water), availability of land, and extant operational facilities (transportation and communication, power, and a nearby civilian community). A short list of five potential test sites was developed, including Alamogordo–White Sands Guided Missile Range in New Mexico (where the Manhattan Project's 1945 Trinity test occurred); Dugway Proving Ground in Utah; the Las Vegas Bombing and Gunnery Range; an area near Fallon, Nevada; and the Camp Lejeune area in eastern North Carolina. Since availability of land was desirable not only to avoid "complicated acquisition processes" but also "from the security and public relations points of view," the latter two sites were dropped from consideration because they were not already under military control.[28]

On December 13, 1950, the AEC recommended to the National Security Council that 5,000 square miles of the Las Vegas Bombing and Gunnery Range be developed as an atomic test site. Nested inside the bombing range, the Nevada Proving Grounds, later renamed the Nevada Test Site (see figure 1.1), had best satisfied the commission's four criteria, including having the fewest people residing in a "90 degree possible fallout sector, to a radius of 125 miles

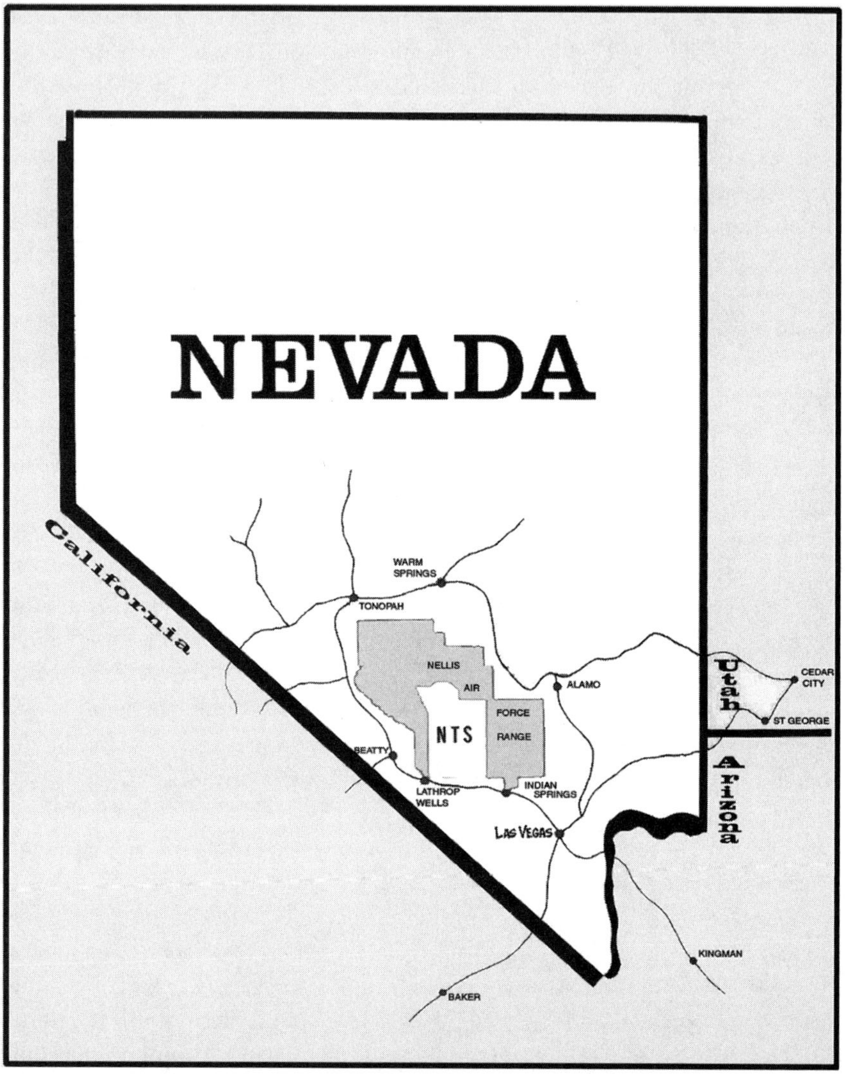

Figure 1.1. The Nevada Test Site (NTS), located sixty-five miles northwest of Las Vegas and nested inside the Nellis Air Force Range, was established in 1951 as a 5,000-square-mile proving grounds for nuclear explosives testing. As reflected in this map, NTS was later expanded to 13,500 square miles. *U.S. Department of Energy photograph P-2808, enhanced by Jun Liang.*

downwind from site." It was also the most convenient for Los Alamos and AEC Albuquerque personnel. According to Los Alamos scientists assessing radiation safety issues for the Nevada site, sparse population and the extensive spaces of the site would provide greater public safety than Alamogordo, New Mexico, had provided for the original Manhattan Project blast. Investigators concluded that a tower-burst bomb of up to 25 kilotons, perhaps even 50 kilotons, could be detonated without exceeding the "allowed emergency tolerance dose" for a 180-degree area extending one hundred miles in radius away from the population center in Las Vegas.[29]

In making the test site, the AEC was also producing another kind of space, the "fallout sector" abutting it. While the Alamogordo site had been used only once, the Nevada Test Site would be used far more frequently, making places downwind susceptible not just to emergency conditions but also to the accumulation of long-living, low-level radiation as well as radiation hot spots from the Nevada tests. The boundaries of the supposed 125-mile fallout quadrant, along with the AEC's confident claims about predictable wind patterns, would prove to be deeply flawed, as not just eastern Nevada and southern Utah but also parts of Idaho and Montana would emerge as environments most contaminated by atmospheric fallout from Nevada tests during the 1950s and early 1960s. Nevertheless, the very act of mapping such boundaries and fallout sectors was part of a discourse of spatial control (working through a calculus of permissible levels of radiation, size of explosive, weather patterns, and so forth) that was, from the beginning, central to AEC testing practices in Nevada and elsewhere.

The public representation of the new proving grounds began with a press release on January 11, 1951: "The use of the Las Vegas Bombing and Gunnery Range will make available to the Los Alamos Scientific Laboratory a readily accessible site for periodic test work with a resultant speed-up in the weapons development program. Test activities at the new site will include experimental nuclear detonation for the development of atomic bombs—so-called 'A-bombs'—carried out under controlled conditions."[30] But just how controlled could the conditions be for experimental nuclear detonations? This question was left largely to the AEC and its scientists, but safety from Nevada testing, for the public and for nearby Air Force personnel, was assured by the AEC and Senator Brien McMahon, now chair of the JCAE.[31] The first test series in Nevada was to begin the following month.

EDWARD TELLER AND THE LIVERMORE LABORATORY

The test site in Nevada, like the Pacific proving grounds before it, was created as a resource for the Los Alamos weapons laboratory and for the stakeholders in its work. Even as the continental test site was being established in Nevada, however, some in the weapons science community were arguing that another laboratory was needed as well. Most vocal among them was Edward Teller (figure 1.2). Born and raised in Budapest, Teller, who had earned his doctorate under Werner Heisenberg at Leipzig in 1930 at the age of twenty-one, eventually managed to leave Europe for the United States and a full professorship at George Washington University in 1935.[32] At Los Alamos, rejecting his assigned task of making calculations for the fission bomb, which was already in essence theoretically feasible, Teller had begun calculations on possibilities for the hydrogen bomb—known as the "super"—which would produce a thermonuclear explosion based on the fusion process (rather than a nuclear explosion based only on the fission process). As Teller later recalled his pursuit of the super at Los Alamos, "This was the kind of a job that was really delightful for me—to try to speculate in these wide ranges."[33] Debates over whether or not a second national nuclear weapons laboratory was necessary were bound up in the scientific and political history of the super, and equally so in the history of Teller's own work as both an entrepreneur and an advocate in the politics of military science and technology.

In public commentary over nearly five decades, Teller complemented his black-and-white precepts of anticommunism and deterrent "peace through strength" with the claim that, in a democracy, it was the job of the scientist chiefly to *demonstrate* "man's power over nature" and that of elected representatives to decide how that science and technology should be used.[34] His own politics belied this point. Although the AEC had, over Teller's protests, put off a high-priority program for the super in 1949, Teller continued to pursue thermonuclear fusion—both in his scientific work and through the forging of political ties. Los Alamos successfully conducted the first small-scale thermonuclear experiment in May 1951, at Enewetak, based on the design of Teller and mathematician Stanislaw Ulam.[35] But Teller's dissatisfaction with the laboratory persisted. By October 31, 1952, when the first large-scale thermonuclear bomb was exploded at Enewetak, Teller had already left Los Alamos to take a position at the University of Chicago, where he remained actively engaged in the project of building a second national weapons laboratory. Teller's conflict with Los Alamos director Norris Bradbury and his senior staff—essentially over the speed at which the super was pursued—had apparently led him to believe that another national weapons laboratory was necessary to vigorously pursue thermonuclear weapons design. A second weapons laboratory, Teller argued, one devoted to the development of new—and perhaps theoretically riskier—ideas, would spur creativity and productivity in both labs, even friendly competition, and help to keep the American nuclear program a step ahead of the Soviet Union in the contest for technological supremacy. Yet

Figure 1.2. Edward Teller in 1958. *Credit must be given to the University of California, Lawrence Livermore National Laboratory, and the Department of Energy (Atomic Energy Commission) under whose auspices the work was performed when this information or a reproduction of it is used.*

Teller had difficulties convincing the GAC, composed partly of senior scientists (including Oppenheimer) loyal to Bradbury and Los Alamos, of this need. They argued that a second laboratory would detract from the first, not least in terms of scientific personnel.

Teller next took his case directly to the air force, which was especially interested in the development of relatively small and light hydrogen bombs fit for its airplanes. Enthusiastic support from David Griggs, chief scientist of the air force and a founder of the RAND Corporation, and James Doolittle, a retired air force general and science advisor, ultimately led to air force sponsorship at the highest levels in Washington.[36] In 1951 the air force began subsidizing some of Teller's research at Chicago. Although this was a small-scale effort, the air force's support of Teller, and its interest in a second laboratory, seems to have been a key turning point. As Herbert York later argued, "However serious their [the air force's] intentions concerning a full-scale second laboratory may have been, their action greatly increased the pressure on the AEC either to do something itself or to see its monopoly in the field eliminated."[37] In addition to the air force, Teller enlisted the support of JCAE chairman McMahon, and was given audiences with the secretary of state, secretary of defense, and others. By late 1951 serious consideration in the AEC was being given not just to whether to establish such the new laboratory, but also to where it should be located.

While still at Los Alamos, Teller (along with other scientists) had farmed out some diagnostic measurement work related to the first thermonuclear experiment by forming—with the help of Ernest Lawrence—a "Measurements Project" composed of some forty young Berkeley scientists. This group had been using space available at a decommissioned World War II naval air station in the town of Livermore, forty miles southeast of Berkeley, for operating and storing experimental equipment, including a large particle accelerator, a de facto extension of Lawrence's University of California Radiation Laboratory. And so Lawrence, who shared with Teller a strong commitment to thermonuclear weapons research, was enlisted into the second laboratory project and together they proposed to establish the new laboratory at Livermore as a new weapons branch of the University of California Radiation Laboratory. The GAC approved the proposal in part because of these extant facilities and personnel at Livermore (and Berkeley), which would allow the new laboratory to be up and running without detracting from Los Alamos. The AEC established the laboratory in June 1952, and by that September, the University of California Radiation Laboratory at Livermore began its work under the directorship of Lawrence's Berkeley protégé York, then just thirty years old.[38] Teller was named to the Scientific Steering Committee, and as York recalled, "because of his obvious special status, was given veto authority over the decisions of the committee but otherwise had no formal authority."[39]

As an AEC laboratory, Livermore would have relatively secure access to the immense resources of the national security state, but at the same time, the laboratory would be subject to at least some level of political (and potentially, democratic) supervision. Like other AEC laboratories, its budget would be channeled through the AEC in Washington, a process which at times turned AEC program managers into key arbiters of funding and contributed to a climate of both competition and cooperation among the AEC laboratories.[40] For Livermore and Los Alamos in particular, however, any competition occurred within a framework of abundance. Although it was perhaps a self-fulfilling prophecy, proponents of the second weapons laboratory were right in saying that it would not detract from the first at Los Alamos. Despite some budgetary hiccups, the total budget for Los Alamos grew from $48.8 million in 1953 to $82.1 million in 1963.

Livermore would grow faster. In 1953 the laboratory had an annual budget of $11.3 million and a staff of 700. By 1955 the staff had grown to 1,500, one-third of whom were scientists and engineers (a ratio above the norm for AEC laboratories), and by 1958, with a staff of over 3,000, the budget reached $38.3 million. And by 1963, with a staff of 5,000 and a $103.7 million budget, Livermore had outgrown Los Alamos.[41] This growth was, of course, consistent with the massive growth in scientific funding as the nuclear arms race and Cold War geopolitical agenda were pushed forward at this time, and in this sense, as the broader story of the Manhattan Project and the "military-industrial-academic complex" during the Cold War reminds us, it is important to recall that an AEC laboratory was not the only place to do advanced weapons research.[42] The making of the second nuclear weapons laboratory at Livermore was in no sense an inevitable reflection of this growth; it had to be made. The new laboratory, like Teller himself, was supported to the degree that it could be made useful to other powerful actors and institutions; in turn, the laboratory accumulated power of its own—the power to organize research programs and to produce new forms of knowledge and technology for transforming the world beyond its boundaries.

Fallout, Public Health, and the Geographies of Scientific Controversy

With the establishment of a continental test site in Nevada, the AEC quickly learned that, while more convenient for scientific and technical personnel and materials, testing on U.S. soil presented a far more difficult task in public relations than the commission had yet experienced in the remote Marshall Islands. For instance, when the AEC kept secret the schedule for Nevada's second test series in the fall of 1951, *Los Angeles Times* reporter Gene Sherman argued that "newsmen were there as representatives of the public, which is being asked to spend millions of dollars for atomic research. The public is entitled to know as much about its purchases as is consistent with the safety of national defense."[43] Similarly, *New York Times* correspondent Gladwin Hill complained: "All this uncertainty has had the desired effect of confusing the Russians . . . it also promises equally to confuse the American public, on whose support the work depends."[44] But did it?

Recognizing that such public relations issues might in fact jeopardize the test program, the AEC developed a more proactive information policy. This effort should not be confused, however, with greater public inclusion in any real sense, since "the continuing goal," as the Santa Fe Operations Office described its public information plan for an upcoming Los Alamos test series, remained "to remove the possibility of unreasoning public reaction from operations determinations and to permit continued use of a continental test site."[45] While it continued to restrict publication of radiation monitoring results from downwind areas in Nevada and Utah—"fallout sectors" that far exceeded the original criterion for a 125-mile, ninety-degree quadrant specified during site selection—the AEC also sought to establish more cordial relations with the national media. For example, it offered the press an on-site view during an April 1952 "open-house" explosion. For this occasion, a new observation point within the site was constructed to accommodate the press (see figure 1.3), designed so that "uncleared observers would not have the opportunity to mingle and talk with cleared observers who would be in possession of restricted data and might inadvertently disclose some facts falling in that category." In response to media queries for more information, it was argued (internally) that "an official release, carefully phrasing and limiting a subject, is better protection for security than to permit unlimited, unofficial reporting."[46] And yet even as some of the initial concern from the fourth estate could be deflected through the mobilization of spectacular demonstrations such as the atomic open house, the AEC's authority over scientific matters was not accepted by all as the sovereign truth, especially not outside its weapons laboratories, test sites, and operations offices.

For the AEC's atomic scientists and strategists, public concern over the health effects of nuclear testing was considered to be largely unfounded and irrational. Questions of radiation safety, they argued, were best left to the experts. But in the classified community of nuclear testing, expertise was defined, at least partly, by affiliation: to be an expert, one required access to "restricted data," and to gain access to restricted data, one had to be working for the AEC, an agency itself charged with *both* expediting and regulating nuclear weapons development. Writing into law what had already been a matter of AEC practice, the Atomic Energy Act of 1954 codified essentially all knowledge produced at the national laboratories as restricted data, at least until it might be declassified or put forward in some form as a public information release.[47] For this form of technocratic expertise to be maintained in the face of resistance, however, was another matter, for while the task taken up by AEC public information programs may have been to accommodate an (external) "public" to "the facts" of the Atomic Age, by the late 1950s, these facts were increasingly under challenge by oppositional scientists who brought their own forms of expertise, if not restricted data, to public debate on the problem of fallout and public health. These "geographies of controversy" constituted another important realm—the spatiality of contested knowledge and knowledge production—into which Plowshare and its nuclear excavation program were to be contextualized.[48]

AEC nuclear tests were never carried out without public safety plans; indeed, the assumption that safety from fallout hazards *could* be planned and accounted for was perhaps the key assumption in the AEC's testing and radiation safety operations. And yet in practice, the development of radiation safety planning and, for that matter, of the radiation sciences often lagged behind

Figure 1.3. "News Knob" overlooking Yucca Flat at the Nevada Test Site, April 22, 1952. *U.S. Department of Energy photograph NV0315875, Nuclear Testing Archive, Las Vegas, Nevada.*

(and reacted to) the experimental work of explosives testing. For example, while AEC scientists and engineers insisted that a properly run experimental program would keep human radiation exposures within "permissible dose" limits, these limits were revised downward repeatedly during the period of atmospheric and underwater testing by the Federal Radiation Council and the AEC's advisory committee to its Division of Biology and Medicine (for example, in

1951, 1955, 1957, 1960, and 1961), in response to new data as well as to new testing practices.[49]

Nuclear explosives testing was a new and somewhat unpredictable science. As Los Alamos director Bradbury assessed predictability issues for the AEC's Santa Fe Operations Office in 1953: "Any devices to be tested should not create uniquely new problems of contamination or blast, and yield alone should be an adequate measure of the problem to be expected in any given proposal. It must, however, be clear that in an experimental operation the yield frequently will be quite unpredictable and only a 'probable' value can be given which may be exceeded or missed entirely."[50] Experimentation with nuclear explosives, and the operational plans for dealing with their hazards, worked in a grammar of probability, not certainty.

It was such a "probable value" that was shattered by the February 1954 "Bravo" test at Bikini Atoll, a colossal 15-megaton explosion that more than doubled the AEC's expected explosive yield. Radioactive fallout from Bravo, the first "deliverable" hydrogen bomb—with a force 750 times that of the Hiroshima explosion—was reported to be "off the scales" at the nearby atolls of Rongelap and Utirik. But AEC radiation safety experts, confident enough in their abilities to predict wind patterns, had removed both of these atolls from Bravo's safety plans, leaving islanders without basic information concerning how to wash fallout off their bodies and their food. While bunkered military personnel in the area were evacuated by plane within thirty-six hours of the blast, it was another day before 82 islanders from Rongelap were evacuated (by boat), and fully three days after Bravo that the boats arrived to evacuate Utirik's 157 inhabitants. In a crude attempt at damage control, AEC chairman Lewis Strauss appeared on the scene and proclaimed that the "natives . . . appear to me to be well and happy."[51] In fact, they were still suffering from the initial symptoms of radiation poisoning: skin lesions and "beta burns," vomiting, hair loss, and other maladies. They were displaced from their homes and fishing waters for three years, and deadly thyroid cancers began to attack their children nine years after Bravo.

The Bravo fiasco attracted a great deal of international criticism to the United States and its AEC, particularly after a Japanese fishing boat (the misnamed *Lucky Dragon*) returned home from the Marshall Islands a week later with a sick crew and a cargo of radioactive tuna. For the AEC, the emerging "fallout panic" was perceived as a threat to testing both in the Pacific and in Nevada, where ranchers were beginning to press claims against the AEC over what they perceived as, but typically lacked sufficient evidence to prove really were, damaged, toxic livestock.[52] Chairman Strauss, for his part, sought to quell the bad publicity by stating that, the Bravo accident aside, the global increase in radiation from nuclear testing was "far below the levels which could be harmful in any way to human beings, animals, or crops."[53]

Strauss's claim, of course, matched well with AEC desires to continue the testing program, but it was also largely in step with extant norms in the physical and medical sciences. As scientific debates over fallout ensued, numerous AEC scientists—from physicists at the weapons laboratories to physicians at the AEC's Division of Biology and Medicine (DBM)—elaborated Strauss's point again and again, arguing essentially that "permissible doses" of low-level radiation were insignificant compared with the amount of radioactivity received through natural and artificial sources like solar heat, medical X-rays, and luminous wristwatches.[54] Many geneticists disagreed. By 1954, according to A. H. Sturtevant of the California Institute of Technology (Caltech), it was "widely confirmed" that high-energy radiation produced mutations in reproductive cells; that the frequency of mutations was proportional to the dosage of radiation; that there was "almost certainly" no "threshold value" below which radiation was insignificant; and that the genetic effects of exposure to radiation were cumulative and permanent.[55] Although geneticists like Sturtevant did not dispute the fact that nuclear testing had, thus far, contributed but a minute fraction of the total human exposure to radiation, they also argued that radiation exposure beyond natural sources or wise medical uses should nevertheless be minimized, since *any* increase of exposure levels brought with it a corresponding increase in the risk of genetic mutations.

The issue became increasingly popularized, and more explicitly politicized, when Adlai Stevenson advocated a ban on thermonuclear weapons testing during his 1956 presidential campaign, prompting some four hundred scientists to weigh in with public statements on either side of the testing controversy. Special committee studies commissioned by the National Academy of Sciences produced mixed, and decidedly controversial, results. The Pathology Committee, composed mainly of medical scientists and chaired by a former director of the DBM, concluded that nuclear test explosions could be increased tenfold "without causing any serious genetic danger." Sturtevant, a member of the same body's Genetic Effects Committee, contested this characterization in a letter to the *Washington Post*.[56] And in a statement reflecting just how far was the Atomic Energy Commission of 1956 from the scientists' movement's vision, a decade earlier, of a publicly accountable government science, twenty-four scientists from Washington University in St. Louis charged that America's atomic policy had been developed "in a vacuum of public information" and, alternatively, called for "intensive scientific study and public discussion" of the fallout hazard and its policy implications. Others, however, like their university colleague and former Manhattan Project scientist Arthur Compton, acknowledged some genetic risk to the population from continued testing but insisted that thermonuclear tests nevertheless remained vital "to maintain our freedom."[57]

Controversies over the genetic effects of radiation (on germ or reproductive cells) were soon matched by disputed claims over the pathological effects of radiation (on somatic cells involved in the formation of bodily organs and tissues), focusing chiefly on the relationship between leukemia and the isotope strontium-90 (Sr-90). Although Sr-90 had not existed prior to the Manhattan Project, it was, by this time, present in small quantities in every plant and animal the world over. Commonly taken up through the ingestion of milk and vegetables, Sr-90, due to its chemical similarity to calcium, tended to accumulate in bone marrow, where it continued to emit radiation. In May 1957, the Nobel Prize–winning geneticist and biological chemist Linus Pauling, also at Caltech, estimated that ten thousand people had likely already died or were dying of leukemia because of nuclear testing. Pauling's estimate was supported by biologist E. B. Lewis, another Caltech colleague, in a quantitative analysis published in *Science*.[58] And yet in the same year the AEC could proclaim, in a pamphlet produced for southern Nevada and Utah residents, that its research had "confirmed that Nevada test fallout has not caused illness or injured the health of anyone living near the test site."[59] Who was right?

The state of scientific and political controversy over the health effects of fallout made for a complex state of affairs, what might be called a state of *normal* controversy. It allowed the AEC to continue its nuclear testing program, for the time being, at its laboratories and test sites. At the same time, normal controversy generated still more oppositional science, and public politics, seeking to end or limit the tests. While the AEC still wanted to remove the threat of "unreasoning" public participation from its operations, scientifically oriented activist organizations like the Citizens for a Sane Nuclear Policy (SANE), Greater St. Louis Citizens' Committee for Nuclear Information (CNI), and the Federation of American Scientists (formerly the atomic scientists' movement and then the Federation of Atomic Scientists) were after just the opposite: they sought to expand public knowledge of fallout and radiation science with the idea that people could then better govern themselves, making judgments on the basis of adequate information and hard science.[60] By February 1957, as ideas about nonmilitary uses of nuclear explosives were being hashed out at Livermore, international calls for a nuclear test ban or moratorium were growing. In June 1957 the Soviet Union itself proposed a two- or three-year moratorium on testing, and with world opinion at stake, the Eisenhower administration was giving the matter serious consideration.[61]

Edward Teller, by this time widely known as the "father of the H-bomb," was already the winner in a number of key debates in America's Cold War politics of science and technology, including, as we have seen, the decision to build the second weapons laboratory as well as the 1954 Oppenheimer affair, in which he provided testimony damaging to his former Manhattan Project laboratory director. Together with AEC chairman Strauss, Ernest Lawrence, and Livermore's Mark Mills, Teller visited Eisenhower in the Oval Office just ten days after the Soviet overture to argue against restrictions on nuclear testing. Teller told the president that a completely "clean bomb," with virtually no fission component and thus no radioactive fallout, could be developed in a "matter of six or seven years" given continued testing.[62] Whatever the influence of Teller's argument, which hindsight tells us was wishful thinking at best concerning technical progress on clean bomb research, Eisenhower delayed U.S. participation in the moratorium until October 1958. Meanwhile the AEC accelerated testing in Nevada: an astonishing twenty-nine nuclear bombs were set off in Nevada between late May and early October of 1957.[63]

Teller's politicking was by no means limited to the role of presidential advisor; as the clamor for an end to nuclear testing continued, he became equally active in public debates over the dangers of fallout. Although many scientists, politicians, activists, and others were involved in this debate, it came to be personified in the popular media in the oppositional figures of Teller and Linus Pauling. In January 1958, Pauling presented a petition to the United Nations that he had coauthored with biologist Barry Commoner and physicist Leo Szilard. The petition, signed by more than nine thousand scientists, called for the cessation of nuclear testing. It affirmed that "each added amount of radiation causes damage to the health of human beings all over the world and causes damage to the pool of human germ plasm such as to lead to an increase in the number of seriously defective children that will be born in future generations."[64]

Pauling's petition was attacked in a 1958 book and corresponding article in *Life*, "The Compelling Need for Nuclear Tests," written by Teller with Dr. Albert Latter of the RAND Corporation.[65] Teller and Latter argued that the threat of radioactive fallout from tests had been exaggerated. "Unfortunately," they lamented, "much of the discussion about continued nuclear experimentation has been carried out in a most emotional and confused manner."[66] With emotion thus expelled from their argument, Teller and Latter insisted that the actual dangers from cancer and leukemia causing radionuclides such as Sr-90 and cesium-137 were minimal for the "average American," amounting to a dosage of only a small fraction of one roentgen per American.[67] "If the tests continue at the present rate, radiation levels might increase as much as fivefold," the physicists conceded in *Life*, yet they maintained that "even in this situation it is extremely unlikely that anyone would receive a lifetime dosage of as much as five roentguns. If radiation in this small amount actually does increase a person's chance of getting bone cancer or leukemia, the increase is so slight it cannot be measured."[68]

Pauling responded in a letter to the editor by challenging many of Teller and Latter's points as either false or misleading. The emphasis on the "average American," for example, obscured the geographical specificity that fallout took. Pauling argued that it was not at all unlikely that lifetime doses would exceed five roentgens, and he cited as evidence an AEC report that listed several towns in Nevada and Utah where thousands of inhabitants had already been exposed to more than five roentgens of fallout from a single test series in Nevada. Teller and Latter countered in a response that they had meant that such lifetime doses were unlikely outside the "immediate downwind area"—they had written off the immediate downwind populations, in other words, as statistically insignificant—and they reiterated their mission to "allay fears aroused by exaggerated and frightening statements."[69]

Next Pauling extended the argument onto the more sympathetic pages of the *Nation*.[70] While he had objected to the spatial scale of resolution in Teller and Latter's analysis of the "average American," in this article he objected to their temporal scale of resolution in reference to the Bravo test. Teller and Latter had claimed that no malignancies or cases of leukemia had been reported to date after Bravo, yet this was misleading: for either bone cancer or leukemia, expected deaths to the islanders would occur within a twenty-five-year time frame, and so observations in just three years' time had no meaning (an argument that could also be applied to the AEC claims of safety to Nevada and Utah residents in its 1957 pamphlet). And yet the controversy over the effects of radioactive fallout was not given closure at this juncture, even after a televised debate between Teller and Pauling.

So who *was* right? The answer depended largely, at this point, on which persons were asked, where they worked, and the nature of their scientific and political backgrounds and perspectives. There was not overwhelming evidence on either side of the debate over how harmful the addition of (relatively) low levels of radiation would be—not overwhelming enough, that is, to overcome these differing, intermingling scientific, institutional, and political perspectives. But for those paying attention to the debate, the controversy must have at least provided a sense of the inseparability of science and politics when it came to matters of nuclear testing, fallout, and public health. For Pauling: "The compelling need in the world today is not to continue tests for nuclear weapons; it is the need to stop wasting world resources on armaments . . . and to bring morality and justice into their proper place of importance in the conduct of world affairs."[71] But for Teller and Latter, the geopolitical context in which the tests took place trumped not only these questions of morality and justice but also the exercise of caution in dealing with the unresolved problem of radiation hazards: "The actual difference of opinion concerning the facts of radiation hazards is not great. To our minds the important issue is whether the small and uncertain hazards of fallout will induce us to abandon one of our most important methods of defense in the face of real danger of Communist aggression."[72] Or, as Teller would refine his argument just two years later, "Most Americans are dangerously confused about the hazards of radioactivity. They fume about test fallout. But this is no danger at all."[73]

Inseparable from the state of normal controversy was thus a politics of uncertainty. As long as scientific truth was uncertain, and open, at least to a degree, to the competing claims of different scientists and institutions, then information could be organized in ways that supported different positions. The still evolving fallout controversy, as a contested relationship among—and not simply between—scientific and political communities, provided the context and rhetorical frame for scientists' "public entries" into the politics of nuclear testing during the late 1950s. At the same time, this condition of controversy in some ways shaped the kinds of research that were conducted at numerous institutional sites, as university geneticists, AEC medical physicists, and National Academy subcommittees all sought to validate their arguments or to adjudicate between differing positions. These differences, and the contested nature of knowledge production about the health effects of fallout, were constituted in space as well as time, a geography of scientific controversy characterized by the simultaneity of different truths in different kinds of scientific spaces.

This is not to say that either side of the controversy existed in isolation from the other because they tended to reside in different geographical spaces; on the contrary, the shared regulatory context of national policies and international agreements was the common ground that testing advocates and opponents were fighting over. Nor did where someone worked ultimately determine his or her stance on the fallout question. In 1956, for example, seventy-three scientists from the AEC's Argonne reactor laboratory signed a declaration, originating from Caltech, that called for the discontinuation of large nuclear tests, which suggests that AEC boundaries did not always serve to fence out what the leadership of that institution likely saw as the wrong kind of politics.[74] But it was Argonne, not Livermore, where these protests occurred, and similarly, it matters that many of the loudest voices raised against testing, and the studies conducted to support them, also came from particular sites, like that of the genetics community at Caltech. When Livermore's Harold Brown acknowledged after what would come to be known as the first Plowshare symposium that, until very low fission explosives were available, large-scale nuclear excavation blasts would be limited to "uninhabited areas," the risks involved, and the problems presented by the movement of radioactivity beyond Brown's de facto safe distance of one hundred miles, meant something quite different to participants in Livermore's "non-military uses" symposium than they would have to those at a genetics colloquium at Caltech or at the Biology Department at Washington University in St. Louis.

This point should not be taken for granted. I have argued that the Plowshare program was born in particular sites of knowledge production that were themselves products of and participants in a wider network, the American military-industrial-scientific complex during World War II and the early Cold War, a network of relations that was organized around demands for the dynamic production of scientific and technical knowledge and artifacts. In contrast to the traditionally "public" spaces of modern science, these worksites of American nuclear explosives science were structured—in some ways, but by no means all—by differing norms, different experimental and analytical practices, and differential access to classified data. The complex mix of scientific, institutional, and political issues bound up in the fallout controversy were thus both reflected and reproduced in the work of places like Livermore, where the notion of an experimental program in nuclear excavation, and nonmilitary uses more generally, seemed like a good idea in February 1957, an idea that was worth at least modest support from the AEC later that year. But an "internalist" genealogy of Plowshare cannot be enough; while it is imperative that we locate the program's original agents inside of Livermore's gates—and not as a mere response "triggered" by outside world events such as the Suez crisis—the development of the program also needs to be understood in terms of a relationship between insides and outsides, the laboratory and the "public," and thus also in terms of the "boundary work" that both limited access to information and underwrote the program builders' claims to expertise.[75] In Plowshare's entry onto the national and international scenes, it was the controversial world outside Livermore's high-security gates that the very idea of the "peaceful nuclear explosion" was intended to shape.

Enter Plowshare

The first significant movement of Project Plowshare from classified space to public space was a rhetorical one. The publicity generated for the program in early 1958, that is, had more to do with effecting a "change in public attitude" toward nuclear testing—as indicated a year earlier at the Livermore symposium—than with "chang[ing] the earth's surface to suit us."[76] But first, the authors of this publicity had to get their terms straight. At a meeting on September 10, 1958, Livermore's Jim Frank, Jim Ruff, and Harlan Zodtner met with AEC public information specialist Ernest Campbell to produce "a glossary of terms for the Plowshare program [that] would be helpful to people writing articles." As the Livermore team saw it, "the use of new terms in industrial applications will probably be helpful in building entirely new conceptions and background feelings regarding the use of nuclear devices."[77] Their suggestions are listed in table 1.1. As will be seen in chapter 2, efforts to learn from previous (explicitly military) tests about the dynamics of nuclear cratering were by this time already under way at Livermore, along with plans for blasting the first nuclear harbor in a remote region; even at this stage, Plowshare was not *only* a matter of rhetoric. But it is notable that the first public glimpses of the program, and the first, most sanguine visions of nuclear excavation's boundless

Table 1.1
Suggested Language for Plowshare versus Military Terms, Livermore, 1958

Present terms—military	***Suggested Plowshare terms***
Bombs—Weapons—Devices—Warheads	Devices—Nuclear Sources Nuclear Assemblies—e.g., "lowering nuclear assembly into the ground" Nuclear unit HE—high explosives NE—nuclear explosives
Clean	Low fission to fusion ratio Small fission yield
Non-military uses	Engineering uses Industrial uses or industrial applications Industrial and Scientific Uses avoid "peaceful uses"
Fallout	Debris—radioactive debris
Nuclear test	Nuclear experiment—application—demonstration—utilization (of a device)
Fire/detonate	Fire—detonate (use interchangeably)
Ground zero	Site—"D.P." "Detonation point"
Yield	Energy release—yield—yields in BTU, megatons, etc.
Costs (for weapons)	Charges (for experiments)

Source: Adapted from Jim Frank, Jim Ruff, and Harlan Zodtner to Distribution, "Memorandum: Glossary of Terms for Plowshare Program," September 18, 1958, Plowshare Box P-29, Lawrence Livermore National Laboratory (LLNL).

potential, came alongside discussions of a possible nuclear testing ban and, more specifically, with calls for pursuing experimentation in "industrial uses" even in the event of a test moratorium.

HOW TO BE AN OPTIMIST IN THE NUCLEAR AGE

Among Plowshare's first public advocates was Dr. Willard Libby, a Nobel Prize–winning physical chemist who was, at the time, the only scientist among the five Atomic Energy commissioners. Testifying before the Senate Subcommittee on Disarmament in March 1958, Libby described a nuclear test conducted in Nevada in September 1957. Libby informed the senators of a (relatively) small, 1.7-kiloton nuclear explosive test, code-named Rainier, which had been detonated 1,900 feet inside of a desert mesa at the Nevada Test Site. When the bomb exploded, Libby explained, the shock wave caused the *mountain* to jump about six inches, but no radioactivity escaped to the surface. Instead, the explosion left a cavity fifty feet deep encased in a skin of fused rock. Libby believed that the experiment, as he suggested the following month in a speech delivered at Amherst College and distributed through the AEC's public information office, had enormous significance for the "non-military or peaceful applications of nuclear explosions, and the possibility that much of the testing can in the future be done under conditions of no radioactive fallout."[78]

Journalists agreed. "Chemical explosives, comparatively puny, have shaped the modern world," observed *Time* magazine, "dug its canals and harbors, won its coal and ore, but until recently, atomic explosives were too secret and too radioactive to be considered seriously. Now it appears that within a few years they may become man's most powerful tool for fitting the earth to his use." *Time*'s science page indulged readers by wondering how the enormous force of atomic explosions "could be put to peaceful use." After mislabeling Rainier as "the world's first tentative trial of peaceful atomic blasting"—Rainier was chiefly an experiment in the feasibility of underground testing itself, and in how underground tests could be detected seismically—*Time* inferred from Libby's testimony that "an atom-blasted harbor, say on the almost harborless coast of Peru, should be safe after a short time."[79] But while Rainier had produced potentially significant data for Plowshare applications such as the stimulation of oil flow or the creation of underground gas storage chambers, its relevance to nuclear excavation explosions, which by definition broke the earth's surface, was minimal. Whatever his intent, Libby's stumping for Plowshare succeeded in blurring the distinction between underground explosions and the idea of nuclear excavation, at least for some. Another reporter enthused: "What underground atomic blasts may be able to do in the future, Dr. Libby could only suggest, but this was breath-taking: they could move mountains, open up new sources of oil wealth and water resources, create harbors, in fact remake the face of the earth and bring incalculable benefits to man—to all mankind, Russians as well as Americans, Communists as well as capitalists."[80]

For Libby, the incalculable benefits that Plowshare was apparently on track to produce fundamentally changed the terms of debate over the test ban; the question of peaceful applications bore "directly on the question of testing." As Libby implored at Amherst:

> Whatever we do agree to, I hope exception will be made of the peaceful uses of nuclear explosions. . . . There is a natural law, I think, which requires us not to turn our backs on nature. If we have chances to discover new things about matter and its behavior, and new ways of doing things, and do not grasp these opportunities, we pay very dearly for committing what is essentially an unnatural act. I think no man or nation can disregard that fact and remain healthy very long. We must proceed with the non-military applications of nuclear explosions under any conditions! Let us all hope these will be conditions of real disarmament fully implemented and inspected.[81]

Such appeals to the "nature" of scientific enterprise, and to the extension of technological control over nature, were made at length as numerous AEC scientists continued to promote Plowshare together with the need for continued testing. Gerald Johnson was quoted in *Popular Science*, using the language of scale to bring out the technological sublime in Plowshare: "In considering possible tasks, the imagination is free to explore projects hundreds or thousands of times larger than have ever been undertaken."[82] Following the October 1958–August 1961 test moratorium—from which, President Eisenhower would unilaterally declare, peaceful nuclear explosives tests were to be exempt—similarly heroic rhetoric would be employed prior to the negotiation of the Limited Test Ban Treaty of 1963, which proposed to ban aboveground and underwater testing but not, as many activists had hoped, to ban testing all together.

Typically, the concerns of geneticists and others over the risks of fallout were sidestepped in these promotional efforts, such as in Gerald Johnson and Harold Brown's 1958 *Scientific American* article which maintained that "appropriate precautions in a given project can reduce [air blast and radiation hazards] to acceptable levels."[83] No evidence was presented to support the Livermore scientists' claims regarding "acceptable levels" of radiation, and despite much discussion and many diagrams of the Rainier test, the paper contained nothing concrete enough for testing opponents to judge whether or not such levels of radioactivity were potentially acceptable to them as well.

The most complete articulation of the idea of Plowshare with the argument against nuclear testing limits would come a few years later from the pen of Edward Teller. Intended for popular audiences, Teller's 1962 *Legacy of Hiroshima* was written expressly to voice the author's misgivings over a comprehensive or even a partial nuclear test ban treaty. In the book, the physicist's love for science, his devotion to nuclear weapons research, his resolve that communism represented a grave threat to the "free world," and his insistence that "limited nuclear warfare" was feasible were all woven into a genial, first-person account. As Teller saw it, "hopes for disarmament and discussions of a test moratorium have again and again postponed needed experiments." Among the most compelling planks in Teller's platform was a chapter on "How to Be an Optimist in the Nuclear Age." In it he described promising developments in the Plowshare program. Building on earlier claims to *Popular Mechanics* readers from an article entitled "We're Going to Work Miracles," Teller described the exciting possibilities that would be enabled by peaceful nuclear explosions, from the "geographical engineering" of harbors and canals to, perhaps someday, using nuclear explosives to modify the climate. "To be an optimist," waxed Teller, "requires courage, but to be an optimist in the nuclear age demands even more: It demands imagination." Enlisting his readers in this exercise of technological and geographical imagining, Teller not only sidestepped but indeed showed a certain contempt for the kinds of precaution that testing opponents were calling for: "Decades should not be wasted in hesitation. We should begin now to make the best use of nature's gifts."[84]

Why the urgency? Teller's confidence in the Plowshare program, and in the idea of "acceptable levels" of fallout, was unflagging, but plainly the timing of his support for Livermore's experimental program, like that of Commissioner Libby, matched well with arguments for continued nuclear testing. In this sense, the idea of the peaceful nuclear explosion could be put to work in efforts to mold public opinion before any earth was actually moved by nuclear dynamite. It is also possible that the nonmilitary uses program was seen as one possible way around a test ban, that is, as a way of supplying military information on explosive phenomenology through so-called peaceful nuclear explosions during any test ban or moratorium. And yet if Plowshare remained only an idea, or only a scheme for defying testing treaties, then its story would essentially end where it began, in a conference room or on the pages of *Time*. But Plowshare was also conceived, from the start, as a prospect for scientific and technical *work*. What else do laboratories do but seek to develop new programs, produce new forms of knowledge and technology, garner new resources and expand their own reach in the process? Plowshare was a new experimental program with growth potential for an enterprising laboratory, one seeking continued expansion (and diversification) at a moment when the enterprise of testing—the material and epistemological backbone of weapons science at

Livermore—was in jeopardy. It was not a matter of either science or politics, either legitimate experimental program or Cold War boondoggle, for Plowshare was all of these things; the various goals of Plowshare scientists, AEC bureaucrats, and other project advocates who advanced the program were by no means incompatible. Besides which, the idea of actually considering and developing earthmoving tasks "hundreds or thousands of times larger than have ever been undertaken" was not, for many of the Livermore scientists, without a certain appeal.

COMING OUT IN GENEVA

In September 1958 Teller and Johnson went to Geneva along with a large AEC contingent for the Second International Conference on Peaceful Uses of Atomic Energy. Among more than two thousand scientific papers presented at the conference, which was attended by scientists from sixty-seven countries, only Teller and Johnson presented work on peaceful nuclear *explosions*.[85] The Livermore scientists also brought a short film to exhibit in Geneva entitled *Industrial Applications of Nuclear Explosives*. The film, set against a simulated, unnamed coastal landscape, showed how an "instant harbor" could be excavated with 1.4 megatons of shallowly buried nuclear explosives.

Outside of the exhibition hall, however, Plowshare received a chilly reception from scientists, politicians, and journalists. As the *New York Times* reported on the proceedings: "During much of the session today the United States delegation attempted to meet objections that had been raised to the United States program to use atomic explosives for peaceful purposes. The program . . . met with private skepticism among Western scientists attending the conference and open hostility from the Soviet delegation."[86] While the Western scientists may have doubted whether the program could be conducted safely, or worried that it might serve to scuttle the one-year test moratorium already scheduled to begin the following month, the Soviets denounced the program on the floor as a plan for illegally acquiring military data. On the global stage in Geneva, Plowshare did not seem to be generating the change of public attitude that its makers had hoped for.

Much of the task of defending Plowshare, an experimental program that had yet to conduct its first experiment, fell to Willard Libby. Libby sought to head off criticism by admitting that confidence in peaceful applications was "not unanimous even in my own country. There are people—and I would point out that in many instances they are people who are less familiar with the actual explosion phenomenology—who do doubt seriously the usefulness [of peaceful nuclear explosions]." Those who questioned the usefulness or, perhaps, the safety of Plowshare, Libby hinted, might lack the knowledge—and the access to restricted data—to speak authoritatively on the matter. While the commission was not certain that all of the peaceful applications mentioned would materialize, Libby added, "there is no doubt at all that we could make a harbor, and that we could make it at an expense which will make it very attractive in certain areas of the world."[87] Yet after this early episode of boundary policing, it must have already been clear to Libby and the Livermore scientists in Geneva that the actual production of a nuclear harbor would be far more difficult than in the movie.

Chapter 2 Toward an "Early and Obvious Demonstration"

THE QUESTION WAS where to begin. As early as September 1957, the nature of the experimental program in nuclear excavation was starting to take shape. Laboratory director Herbert York had requested that at least sixty-five Livermore scientists, engineers, technicians, and other workers be assigned to Plowshare for fiscal year 1958, including approximately eighteen "senior people," but how would their energies be best spent?[1] As Harold Brown had summarized after the February symposium, "mocked-up tests" (using chemical high explosives) and "paper studies" would be relatively inexpensive, but of little value without "relevant experimental data" from actual nuclear tests.[2] The goal, as Brown and Gerald Johnson had resolved by November 1957, was therefore to establish Plowshare as an active program, beyond the culling of data from previous tests, by executing the world's first experiment in nuclear earthmoving. In fact, it was to be more than an experiment. Within eighteen months' time, the Livermore scientists proposed, the Plowshare program ought to design and carry out a "specific demonstration of a useful excavation."[3] But where?

Considering that a spatial, technological, and security infrastructure for nuclear testing already existed in both Nevada and the Marshall Islands, either of these sites would have been an obvious place to conduct such a test. Instead, Plowshare scientists looked elsewhere—from the Canadian Arctic to Alaska to northern Chile to Christmas Island in the Pacific—for sites where, they believed, a more compelling demonstration of the practical uses of nuclear bombs could be produced than was possible at either test site. In this respect Brown and Johnson's use of the term *demonstration* is telling, implying that they planned to publicly show a working experiment rather than to privately "try" an experiment in which expectations remained somewhat uncertain.[4] Although nothing quite like this had ever before been attempted, Plowshare scientists believed that their experience with nuclear testing in Nevada and the Pacific, along with cratering data extrapolated from high explosives tests, would provide them with sufficient knowledge to demonstrate a "useful nuclear excavation" at this point. Either that, or they believed that they could both try and show their experiment at the same time.

It was for both technical and rhetorical purposes that Plowshare scientists wanted to carry out the experiment outside the boundaries of AEC test sites. On the one hand, geographical variations in the physical and biological conditions of particular places-to-be-engineered made it desirable, from the standpoint of developing a practical nuclear excavation technology, to experiment in varied soil media, topographical characteristics, and biological environments. On the other hand, there was a perceived need to demonstrate the utility of geographical engineering in geography "itself," and not, that is, at the established nuclear proving grounds, already firmly associated with radioactivity and the business of weapons testing. And yet this strategy of simultaneously trying and showing an experiment in nuclear excavation gave rise to new problems, for there were few places where a useful demonstration (that is, a functional harbor or canal) could be made that were actually uninhabited.

The solution, according to the Plowshare physicists–cum–regional development planners, was to go north: Alaska. Above the Arctic Circle, located on Cape Thompson on the northwest Alaskan coast, distant Ogotoruk Creek was selected as the site of their first "instant-harbor" scheme, Project Chariot. By examining why the experimental program moved to Alaska, and what it hoped to gain from this move, this chapter explores how the interrelated aims of Chariot—of trying and showing, experiment and demonstration—took shape around the social and geographical boundaries of experimental space that Livermore sought to establish at Cape Thompson. How did the language of experimentation, demonstration, utility, and economy shift in response to different audiences in Alaska and elsewhere? How was Chariot promoted, how were its value and its safety to be evaluated, and how was it contested? By moving from the highly controlled spaces of Livermore and Nevada to a more public space, albeit a remote one, Livermore scientists would find their authority challenged on a number of grounds.

Locating a Useful Demonstration

SITE SELECTION

Livermore's site selection study brought together scientists and engineers working under the Army Corps of Engineers, U.S. Geological Survey, Radiation Biology Laboratory at the University of Washington, and the mining division of the E. J. Longyear Company, a Minnesota-based subcontractor. For the organizers of the study at Livermore, the process was a simple matter of scanning the planet for sites that could be deemed suitable for an experiment in geographical engineering. "Presently," Brown and Johnson reported in November 1957, "data are being assembled concerning the locations of interest on meteorology, topography, economic or military value, populations, fish and game problems, fallout, and logistics."[5] But if, at the symposium in February, the expressed consensus had been that, given existing fission-to-fusion ratios, excavation projects would be limited to uninhabited areas, the ensuing search for a useful demonstration seemed to have shifted this criterion to one, as in Nevada and the Pacific, where the population was at least of no statistical significance. Areas under consideration in November included sites along the coasts of Chile, northern Alaska, and Christmas Island for harbor excavations, and the Alaskan Aleutian Islands or the Boothia Peninsula on Canada's Arctic coast for navigation canals.[6]

The geographical extent of Livermore's search was, in a sense, suggestive of the sense of power that laboratory scientists envisaged for themselves: the power to search the globe, or at least much of the western hemisphere, to seek out particular local qualities—to assemble place-specific data—that could be used to determine an appropriate site for Livermore's scientific colony. One Livermore study, for example, made a preliminary investigation of thirty-three different ports on the west coast of South America to examine the "feasibility" of blasting a nuclear harbor in the region. The study settled on Arica, a town of thirty thousand people in northern Chile, as the best candidate for eight primary reasons:

1. This site offers opportunity for major harbor improvement.
2. The volume of trade seems to justify harbor development.
3. This area of Chile has a low population density, there is no vegetation or wild life.
4. Three countries are served.
5. A Free Zone is being established in this area.
6. Arica is on the Pan American Highway.
7. Meteorological conditions seem favorable.
8. There are no islands to the west.

Arguing that fallout could be safely blown out to sea, pending further investigations of the local fishing industry, and that, in any event, there was very sparse population "back of the town," the report concluded that "there is a need for a harbor in Arica." On the basis of the site selection study (which seems to have included neither a visit to Chile nor consultation with anyone from there), a nuclear harbor project indeed appeared feasible. The paper included a series of highly abstract, preliminary diagrams of the spatial layout of nuclear explosives for three-, six-, or twelve-ship harbors. With designs ranging from 10 to 100 kilotons in explosive yields, it was estimated that harbor construction for the Arica project, exclusive of the expenses for the nuclear explosives themselves, would cost between twelve million and twenty-one million dollars.[7]

Whatever power was realized in the ability to conduct a planetary search for nuclear excavation sites, or by the force of such analysis, it meant little without the power to actually set the project in place. By early 1958 it was clear that Arica would have to wait for its new harbor facilities, as the global reach of site selection had been scaled back to sovereign U.S. territory. Edward Teller explained to the AEC's San Francisco Operations Office: "Projects outside the United States introduce further complexity into an operation and would probably have to await one or two successful excavation demonstrations [in the U.S.]." Seeking to "provide an early demonstration of Plowshare capabilities," with the utility of the project to be "apparent and immediate," and shifting symposium guidelines from an "uninhabited area" to a "sparsely populated area so as few people as possible will be affected," Teller, beginning a two-year stint as Livermore's director, turned instead to the resource frontier of Alaska.[8] Whether the national scale limitations on site selection were self-imposed by the laboratory or imparted by government confidants, or if, perhaps, Alaska had been the primary target all along, are questions that remain unanswered. What must have been clear to Teller and his colleagues, however, as they prepared to initiate their geographical engineering program, was that the northern exclave of Alaska, scheduled for statehood January 1, 1959, was a region both internal and external to the United States in important ways—culturally, politically, and, not least, geographically—and thus, especially on its Arctic fringe, a seemingly ideal place for Livermore's first nuclear earth-moving experiment.

On February 10, 1958, the laboratory requested that the E. J. Longyear Company "perform a study of the Economic Mineral Potential of Alaska for the region between Point Barrow and Nome. A long term economic view point should be used," it was specified, "to suggest one or several locations, with reasons therefore, where a deep water harbor would be feasible." Instructed to estimate potential harbor traffic twenty-five years into the future on the basis of mineral potential, but given the caveat that "travel to the areas under consideration is not authorized," Longyear settled on the Cape Thompson site as its first choice for the harbor location (see figure 2.1). Although there was no existing transportation or communication infrastructure in the

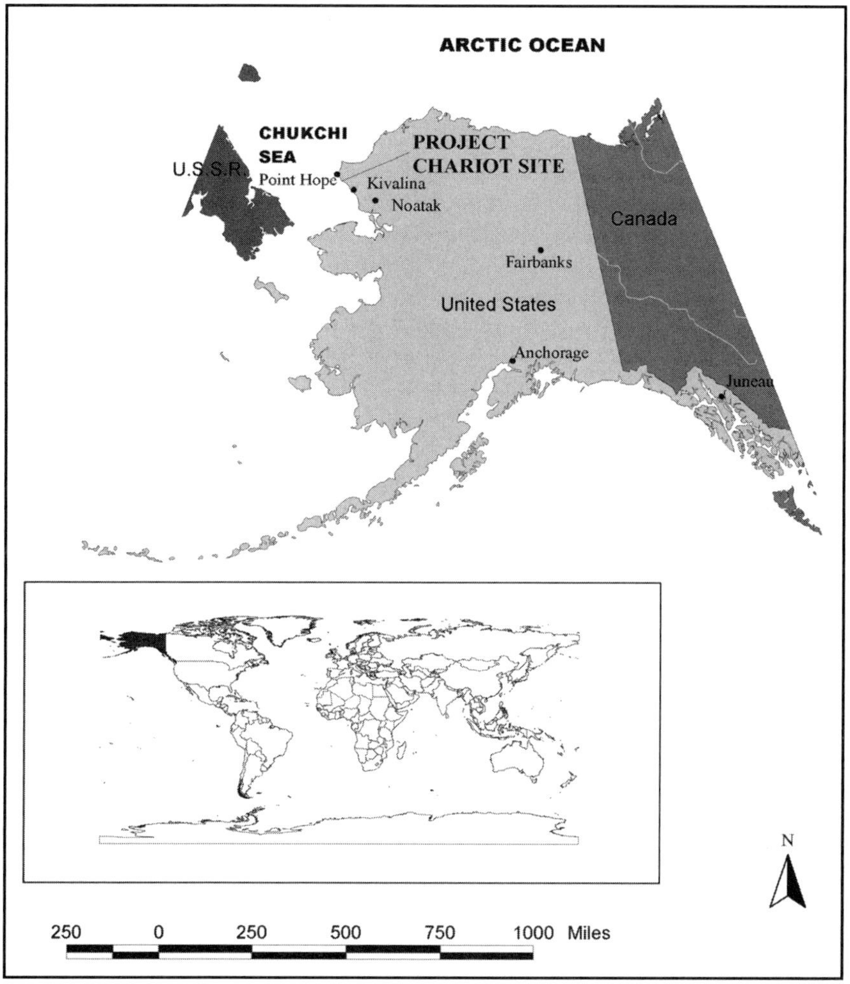

Figure 2.1. Ground zero. The mouth of Ogotoruk Creek on Cape Thompson in northwestern Alaska, just thirty-one miles from the Inupiat village of Point Hope, was selected as the site for Plowshare's first nuclear earthmoving experiment, Project Chariot. *Map by Jun Liang.*

region, Longyear's report argued that the development of railways, roads, and pipelines would be less expensive there because they could largely be built on solid rock, rather than on the "wet" permafrost that characterized much of northwestern Alaska, and would thus obviate the need for more expensive foundation construction.[9] The Cape Thompson site also satisfied two of the laboratory's key site-selection criteria—albeit criteria that were rarely spoken of publicly: remoteness from population centers, and on land owned by the federal government. On May 6, less than three weeks after the Longyear report had been submitted to Livermore, the DMA's General Alfred Starbird reported to the GAC that an initial site for the excavation project had been selected. Livermore requested on May 20 that the AEC withdraw the land from the public domain, where it rested under the control of the U.S. Department of Interior's Bureau of Land Management, asking further "that the above land withdrawal be made as quickly as possible so as to free the area for an early summer survey." The Project Chariot nuclear harbor demonstration, located within forty miles of two Inupiat (Eskimo) villages, was also to be made as quickly as possible.[10]

INTRODUCING CHARIOT: INVENTION AS THE MOTHER OF NECESSITY

"The absence of harbors on the northwest coast of Alaska close to important large-scale mineral deposits," the AEC announced on June 9, 1958, "has in the past hampered development of such deposits. Fishing in this area has also been impeded by the lack of a safe haven." To make this safe haven and to facilitate access to these mineral deposits, the AEC announced that it was investigating the possibility of blasting a harbor in the region near Cape Thompson with nuclear explosives. Personnel from the University of California Radiation Laboratory at Livermore would visit Alaska during the summer "to obtain detailed cost, topographical, geological, and oceanographic data, as well as to ensure that such an operation would be safe with respect to persons within and outside the immediate area and to fish and wildlife." The economic aspects of the project, it seemed from the announcement, were crucial to its inception, necessity the mother of invention. Indeed, when the AEC had approved the decision to move forward with Project Chariot in May 1958, four months before Plowshare's emergence on the international scene at Geneva, it had done so under the impression that there were valuable coal and oil deposits in the region. It had thus authorized the project "provided that the actual execution of the experiment is contingent upon receiving outside sponsorship for development of harbor facilities." The AEC would pay for the experiment—including the explosives as well as "preshot" and "postshot" studies—but someone else would have to foot the bill for turning the "nuclear crater" into a working harbor.[11]

One month after the initial press release, Livermore dispatched a team of scientists to Alaska. Teller and Johnson led the contingent into the territorial capital of Juneau and then on to Anchorage and Fairbanks to promote Chariot to local media, territorial officials, and business and civic leaders. With Alaska's territorial legislature out of session at the time, press conferences and

Rotary Club meetings were quickly arranged to peddle the project to Alaska's pro-development community. In announcing their plans for Project Chariot to Alaska, Livermore scientists were careful to stress the economic benefits of the instant harbor.[12] As Johnson announced in Juneau, "The economic aspect is vital to our planning. We don't just want a hole in the ground. We want to show the Nation what nuclear power can do when used for peaceful purposes."[13]

But while the Livermore team found the Alaskan booster community to be receptive to the idea of nuclear earthmoving, and apparently comfortable with scientists' assurances that the blasts could be accomplished safely or would not be done at all, the economic aspects of Chariot seemed questionable to the Alaskans. Why did the AEC want to make a harbor *there?* A harbor on Ogotoruk Creek would be ice-locked for at least nine months out of the year, and the "proven" mineral deposits that the AEC had described were in fact located hundreds of miles away, across the rugged Brooks Range. To make a working harbor at Cape Thompson would require not only the construction of harbor facilities—estimated by Teller at a cut-rate five million dollars, but by the Longyear Company at from eighty-three to ninety-two million dollars—but also the expense of perhaps a hundred million dollars for the construction of 250 miles of new rail lines across Arctic terrain.[14] A nuclear-blasted canal across the Alaska Peninsula, a dam on the Yukon River, or other nuclear harbors at more economically viable locations were all suggested as alternatives to the Cape Thompson project.

Teller could not have been overly disappointed by these suggestions, for they reflected precisely the kind of geographical imaginations he had hoped to inspire. Enlisting allies for Plowshare from Alaska's pro-development community, these responses indicated, would be easy. By the time the Livermore entourage reached the faculty lounge at the University of Alaska, Fairbanks, where they had been invited to discuss the project with faculty, university officials, and local civic leaders, Teller was openly soliciting suggestions as to where *else* Alaska might be reshaped with nuclear dynamite, even boasting that he and his colleagues could carve a harbor in the shape of a polar bear, so complete was their control over nature. But this sort of bravado would take him only so far among university scientists. Attuned to contemporary debates over the fallout hazard, some of the University of Alaska faculty, especially from the biology department, were not yet convinced that Chariot could be conducted safely, or that the risks posed by the project were as negligible as the weapons scientists had made them seem. Teller brushed aside their concerns with what were becoming his standard arguments: in the first place, the threat of fallout had been greatly exaggerated; second, the amount of radiation that Chariot would add to the atmosphere and biosphere would be far less than the amount that people were already exposed to from background radiation fallout; and finally, the fallout could be controlled in excavation shots through new methods already under development by Livermore scientists.

Whatever weight the Fairbanks scientists gave to these explanations, local newspapers were quick to agree with Teller that radiation hazards were "nonsense." One Fairbanks editorial noted, "Alaska has been invited by the high echelon of nuclear scientists of our nation to furnish the site. . . . We say to Dr. Teller and his associates: Come ahead, Alaska will be proud to be the scene." Why blow the opportunity "for Alaska to have a $5 million dollar earth-moving job at no cost to the new state"?[15] But who would pick up the costs? The biologists were less easily swayed than newspaper editors by Teller's claims about the safety of Chariot, but they may have been relieved that, while the Livermore scientists had failed to attract investors for Chariot's harbor facilities during their trip to Alaska, they continued to sell Chariot in terms of economic development. If the AEC stuck to its position that the project would not go forward if it could not be justified economically, then Chariot would certainly collapse under the weight of careful economic scrutiny.

Meanwhile, as Teller was collecting suggestions in Fairbanks for the geographical engineering of Alaska, Johnson and a crew of scientists and engineers had already traveled to the project site at Cape Thompson. The purpose of this trip was not to reexamine the economic viability of the project, nor was it to inform the local Inupiat communities about the nuclear harbor plan. Rather, the purpose was a three-day "field conference" that included personnel from the AEC, Livermore, Sandia, the U.S. Geological Survey, the Army Corps of Engineers, and the Holmes and Narver engineering firm.[16] There it was settled that Ogotoruk Creek was, in terms of the geological and engineering requirements of the project, the best possible site for a harbor along a twenty-mile stretch around Cape Thompson. Before leaving the area, the team determined where the holes would likely be drilled for placement of the explosives—ground zero—located some thirty-one miles from 317 people living in the village of Point Hope, forty miles from 146 people in Kivalina, and seventy-five miles from 270 people in Noatak. Geological and oceanographic studies related to the mechanics of the harbor excavation were initiated and would continue throughout the summer.

REVISIONS

Although the Livermore scientists spoke confidently of their ability to contain and control fallout, the choice of the location in northwestern Alaska was plainly, from their perspective, a conservative one: Ogotoruk Creek was one of the most remote coastal sites in North America that was (somewhat) suited to harbor construction. And yet the choice also reflected great self-confidence on the part of project scientists in their ability to predict weather and fallout distribution, for Cape Thompson was not remote from the Soviet Union, a fact that could hardly have escaped the AEC's attention, and it is difficult to imagine what the AEC, or the United States, might expect to gain from the possibility of blowing fallout over Soviet territory at this time.[17]

By the time the Livermore contingent departed Alaska on July 19, it was clear that the project could not survive on its thinly supported economic rationale, and that outside investment in Chariot, the precondition for the project set by the GAC, would not be forthcoming. On July 22, when Johnson reported the progress on Chariot to the GAC at Livermore, the primary purpose of the project as a scientific demonstration was emphasized, while the economic arguments for the nuclear harbor—so crucial a week earlier in Juneau—were downplayed. "The possibility of doing a shot that would demonstrate the usefulness of PLOWSHARE would be interesting not only technically, but also psychologically," Johnson argued. "After consideration of many ideas, it was decided that the digging of a deep harbor in northern Alaska would be not too difficult an operation and would be an area in which a change of topography would be useful and impressive."[18]

Johnson also informed the GAC of the latest technical considerations for Chariot. After the field conference, new knowledge of the geology and topography of the area was incorporated into Livermore's harbor design: "Here [Ogotoruk Creek] a 40-foot deep contour runs about 2000 feet from a beach. The plan would be to run a channel 100–300 feet deep, 400 yards wide, 2000 feet long into the location of the present beach—ending in a turning basin. The channel could be cut with five simultaneous 'clean' explosions, three at 150 kt [kiloton] and two at 1.3 mt [megaton]. The excavation would have a volume of about 70 million cubic yards."[19] This original architecture of the explosive layout for the first Plowshare demonstration, then, called for more than 3 megatons of explosive energy. Although Johnson reiterated claims about the mining potential of the region, he concluded his report by admitting that no formal interest in the harbor facilities had been expressed in Alaska.

Early difficulties in selling the harbor to the Alaskans on an economic basis had done little to dampen the enthusiasm of Livermore scientists for the Cape Thompson site. Teller, in an August 15 letter to the DMA's General Starbird, said that "for a demonstration or experiment the site is ideal." Still, Teller was concerned that no one in Alaska could justify the harbor on an economic or military basis. Reflecting an important shift in the promotional and experimental strategies for Chariot, Teller floated the suggestion, attributed to GAC members at the July 22 briefing, of "doing a partial experiment, that is firing only one or two shots to get scaling information."[20] By September plans were under way at Livermore for a scaled-down version of the instant harbor that would no longer be expected to be economically viable, but rather would now be seen only as a "demonstration and experiment" in nuclear excavation.[21]

Teller outlined for the AEC the new rationale for Chariot:

> (a) To date, inspection of the site has indicated that it is ideal from a technical point of view.
> (b) A harbor is the simplest kind of excavation, yet . . . would show that an excavation of some complication could be carried out.
> (c) The project could be done quickly (during the summer of 1960) and would provide an early and obvious demonstration.
> (d) The scaling laws for buried shots in media other than Nevada soil and Pacific coral must be learned. . . . This kind of information, and that on the distribution of the radioactive debris from deeply buried large yield explosions, will be invaluable in planning more complex operations.
> (e) Existing nuclear devices could be used with only slight modification. . . . They would demonstrate the utility of thermonuclear explosives with low fission-to-fusion ratios.[22]

Now Livermore *did* just want a hole in the ground, a *model* harbor from which useful experimental data could be extracted, while still providing a valid public demonstration, helping to pave the way, both technically and politically, for larger, more practical projects elsewhere, including the nascent Central American canal project.[23] Teller argued in a more formal proposal to the AEC in January 1959 that though some cratering data were available from previous (non-Plowshare) tests in Nevada, it was "essential that these indications be confirmed for large scale explosions." By revising the Chariot concept downward, an experiment could still be conducted in Alaska to provide data on the "cratering characteristics of large buried explosions fired simultaneously, the effectiveness of containment or localization of radioactivity, and the suppression of blast effects by burial of the explosive at appropriate depths." The new Chariot design would produce a channel from a "row-charge" of three 20-kiloton explosions and a turning basin blasted by two 200-kiloton explosions (see figure 2.2). "Studies to assure the public health and safety," Teller urged, "need to be initiated immediately."[24] The choice of words was telling, for how exactly could public health and safety be *assured* for what was, by definition—see (d) above—an experiment in the distribution of radioactive debris from deeply buried large-yield nuclear explosions?

Livermore's revisions to Project Chariot were received enthusiastically by the AEC. By March 1959, although the test moratorium had been under way

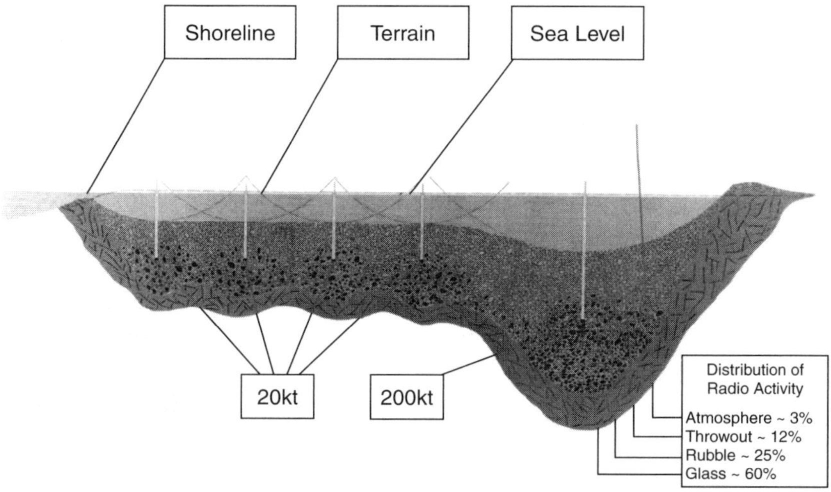

Figure 2.2. Schematic design for Project Chariot. The final configuration for the experiment was reduced again from 460 to 280 kilotons of energy, as pictured here, including a single 200-kiloton crater and a "row-charge" of four 20-kiloton explosives. *Credit must be given to the University of California, Lawrence Livermore National Laboratory, and the Department of Energy (Atomic Energy Commission) under whose auspices the work was performed when this information or a reproduction of it is used.*

for more than four months, the AEC concurred that "a large-scale *experiment* such as that originally proposed for Project Chariot is a necessity before we can proceed with other excavation projects near populated areas," and authorized Livermore to proceed with its studies and plans, adding the caveat to the enterprising laboratory that further consultations would be necessary before any final decisions on the project could be made.[25]

The Project Chariot Environmental Program

In early January 1959, shortly before Teller's revisions for Chariot were submitted to the AEC, Livermore scientists Harry Keller and Vay Shelton had been sent to Alaska. As they stated to chambers of commerce and Rotary and Kiwanis clubs in Anchorage and Fairbanks, Keller and Shelton were there to enlist support for the revised Chariot plan—Chariot, that is, now conceived explicitly as an experiment that would demonstrate the efficacy of nuclear earthmoving techniques. Reassuring Alaska's businessmen that the harbor could be blasted safely, Keller called for resolutions from chambers of commerce and the new state's legislature as a means of showing local support for

Chariot.[26] Even as a demonstration of development, and not development itself, the new scaled-down Chariot still called for the infusion of three million federal construction dollars into the economy of the new state, and local endorsements of the project soon followed from newspapers, chambers of commerce, labor unions, and the state legislature. The *Fairbanks Daily News-Miner* editorialized, "What objections could there possibly be to this large-scale atomic harbor-blasting project? . . . Scientists who have studied the entire matter have given assurances. . . . Is the harbor necessary? It is—and vitally necessary—if the program of peaceful uses of atomic power is to go forward. This harbor blast will be the forerunner of experiments which will lead to the widespread use of nuclear energy in large construction projects. Alaska's 'Yes' vote on this vital project will result in incalculable benefits to all mankind."[27]

Alaska senator E. L. "Bob" Bartlett, after some initial concern, also expressed support for Chariot following a conversation with AEC commissioner Libby and "some folks from Livermore." The scientists, Bartlett commented, "have convinced me they can do a marvelous construction job with exceedingly slight danger to life of any kind."[28] But while the new senator was easily swayed by celebrity scientists like Libby, fundamental objections to the project emerged when Keller and Shelton agreed to speak with the Fairbanks faculty on January 9. Many of the University of Alaska biologists were beginning to question whether Livermore's Plowshare team was actually competent to "study the entire matter."

If university scientists had been relieved by Livermore's insistence, during its first tour of Alaska, that its harbor plan would not be pursued unless it was economically viable, then the changes in the rationale for the project expressed during the second tour, from the economic to the scientific, signaled the need for more serious concern. At the same time, this move to the language of experiment also indicated that the grounds of debate had shifted to terrain on which many specialists in Arctic biology and ecology had firmer standing. The Fairbanks scientists thus began to challenge the boundaries of experimental space that Livermore was erecting around Project Chariot. The most controversial matter was fallout. Biologists were vexed that Keller and Shelton could assert, with no background in either radiobiology or Arctic ecosystems and no evidence presented regarding the nature and quantity of radioactivity to be produced by the blasts, that there would be no biological costs from the experiment. The Livermore scientists had claimed that the project would provide useful bioenvironmental data that could be applied in future projects. But it seemed to the Fairbanks scientists, who had already sent a letter to Livermore in October 1958 through their university president Ernest Patty, expressing their interest in bioenvironmental surveys in connection with Chariot (but received no response), that no serious "preshot" environmental studies had been proposed. Debate over radiation hazards and the possibilities of environmental damage from Chariot continued unabated until it was agreed to break and return for more discussion the next day.

On January 10, fisheries biology instructor Tom English read a statement to Keller and Shelton that had been signed by most of the members of the biology and wildlife departments:

> As biologists we can view the proposed explosions (Chariot) as an experiment in engineering with biological side effects. We are told that the results will be used in planning further explosions. It therefore seems especially necessary to gain as much biological information as possible, and we feel embarrassed that it seems necessary to reiterate that useful conclusions are most likely to follow from a carefully considered experimental design. We have been given no assurance that such a study is contemplated at a realistic scale.
>
> We feel neither competent nor obliged to outline in detail an experimental design. Rather, we urge that employees and consultants of the A.E.C. who are familiar with such problems to be assigned to prepare the experimental design. . . . Only after that work has been done can the local biological scientists evaluate the program and bring to bear their special knowledge of Alaska.

The logic of the statement was clear: only by developing a rigorously planned experiment could the AEC actually gain any useful bioenvironmental information from the Chariot blasts; and only by working with specialists in the field of Arctic biology, it was implied, could this information be evaluated with scientific validity. The statement concluded by expressing concern

> about the thinking involved in the justification of the proposed explosion. We have been subjected to conflicting and crudely misleading statements about the possible biological affects of this explosion and the entire weapons testing program by the A.E.C. We are aware that this is not a weapon, but are equally aware that the explosion can only add to the present contamination in the atmosphere. We prefer to consider the contamination problem elsewhere, but desire to record our apprehension at this time.[29]

At the suggestion of Keller and Shelton, the statement was forwarded to Gerald Johnson at Livermore. English added in his cover letter, "the University is willing and even anxious to have its people help you with their special knowledge of Alaska. The biologists certainly stand ready to do so."[30]

Rather deftly, the statement and letter insinuated the Fairbanks biologists into the experiment and perhaps (and this could hardly have been lost on the biologists) into substantial research contracts. If Livermore still hoped to "assure the public health and safety," as Teller had put it, it would have little choice but to incorporate these Alaska scientists into Chariot in some manner, for in its promotional efforts for Plowshare, the laboratory was in the business of enlisting the cultural authority of science, not undermining it by ignoring the concerns of local experts. But whether Arctic specialists would serve Livermore as producers of useful "biological information" for subsequent nuclear excavation projects, or if, as the biologists' statement had suggested, those with expert knowledge of the local ecosystem would be in a position to "evaluate the program and bring to bear their special knowledge of Alaska," was another matter.

EXPERIMENTS IN ECOLOGY

The AEC was not a newcomer to bioenvironmental research; in fact, it was quickly becoming one of the major sponsors of professional ecology in the United States.[31] The AEC's Division of Biology and Medicine supported a number of research programs at the national laboratories, many involving field studies carried out in the "buffer zones" of converted farmlands, forests, and riparian areas surrounding its nuclear reactor facilities. Although much of this work had an applied character in efforts to manage the new problems of radiation release into the environment and to better understand how particular species or ecosystems responded to high- and low-level releases of radioactivity, much was highly theoretical as well. The latter category included brothers Eugene and Howard Odum's pioneering study of energy flows in the coral reef ecosystem at Enewetak Atoll in 1954, and Eugene's ongoing field studies in the environs of the AEC's new Savannah River tritium and plutonium production plant, located outside of Aiken, South Carolina. As historian Joel Hagen writes, the development of atomic energy "provided ecologists with an exciting new set of tools, techniques, and research opportunities." And yet, Hagen continues, and as Eugene Odum recognized, "for the professional ecologist of the postwar period, nuclear power appeared to be a kind of double-edged sword: capable of wreaking environmental havoc, but also capable of unlocking many of nature's secrets for human benefit."[32]

One of the most exciting additions to the ecologist's toolkit was the radioactive "tracer." Tracers had previously been used by physiologists and biologists to explore the metabolism of (individual) plants and animals, but it was not until after World War II—when nuclear reactors provided plentiful supplies of radioisotopes, on the one hand, and the nuclear reservations outside of reactor areas at Savannah River, Oak Ridge, Hanford, and Brookhaven provided relatively natural spaces with which to experiment, on the other, that tracers were used extensively to study the "metabolism" of whole ecosystems. Tracers could be utilized by ecologists as a means of quantifying the movement of matter and energy through ecosystems, and as such, experimentation in ecology—turning the landscape itself into a working laboratory space—became a source of epistemological currency for ecology. Tracers could be used to produce new knowledge of the internal dynamics of ecosystems, including the ability to isolate individual food chains within an ecosystem. The relationship between the AEC and ecology, as Hagen shows, was thus complex and ambiguous; the AEC provided ecologists with new tools, laboratories, and sources of funds, and at the same time, it produced many of the new problems that ecosystems ecologists had begun to investigate.

In 1955 Ohio State University ecologist John Wolfe took an administrative position in the DBM, and in 1958 became chief of its new Environmental Sciences Division. This move reflected an increase in AEC funding for ecology both at the national laboratories and in university-based research. And so it is not surprising that the AEC had in fact been considering at least some manner of pre- and postshot ecological fieldwork in connection with Chariot, to be directed by members of its Environmental Sciences Division and its associated contractors (although Keller and Shelton's apparent ignorance of such plans, when they went to Fairbanks in January 1959 to sell the harbor-as-experiment project, reflects a surprising lack of communication between Livermore and the AEC). Responding to statements made by Wolfe and others, the Advisory Committee to the DBM had suggested in October 1958 that Chariot could produce useful experimental data for ecology as well as for the engineering aspects of nuclear excavation. The "biologic implications of Project Chariot require ecologic surveys well in advance of any operations" the Advisory Committee had noted, urging "that the opportunities to study the ecology in relation to whatever degree of contamination results should not be lost."[33] In this sense, Chariot could be thought of as a tracer experiment writ large. While the practical side of this work could be used to inform the sorts of questions that Teller had noted in his outline of the experiment, dealing with the distribution of radioactive debris from large-scale nuclear excavation, pre- and postshot ecological surveys could also be used to learn more about how energy moved through the Arctic ecosystem and about how that ecosystem responded to the stresses of "geographical engineering."

The Fairbanks biologists could usefully be put to work to these ends. And yet there were clear differences between the Advisory Committee's interest in ecology "in relation to *whatever* degree of contamination results," on the one hand, and the Alaskan scientists' expressed reservations about adding to the contamination already in the atmosphere, on the other.[34] If this disparity concerned Gerald Johnson, however, he did not let on when he responded to Tom English's letter, stating that the Livermore scientists were "most anxious to join with you in the necessary biological studies." The AEC's San Francisco Operations Office, Johnson explained, was establishing "studies involving both the environmental background of the Cape Thompson area, and an evaluation of the radiological effects to be expected from the experiment." Giving some indication of the time frame that was anticipated for the studies, and for Chariot as a whole, Johnson added that "this work will need to be presented in the coming summer." "Thank you for your considered thoughts," the letter concluded, "and I want to assure you and your colleagues that we will need and use all the help we can get, and particularly from those of you who have devoted time and effort specifically to Alaskan problems."[35] By mid-February, Johnson alerted University of Alaska president Ernest Patty that the biological program was being worked out and that, in Patty's words, Johnson would be in Fairbanks at the end of the month "ready to proceed on the spot with the awarding of contracts."[36]

A substantial role in an AEC study was no small achievement for the University of Alaska, Fairbanks, which at that time consisted of some eighty faculty and eight hundred students.[37] Although it would actually take the biologists until late March to put together a formal proposal, the issues raised in the meetings with Keller and Shelton seemed to have been resolved, at least for the time being, by enrolling the biologists in the preshot studies. The aims of the studies were extensive; a complete regional ecology must include observation of

> what plants and animals occur in this region, what their distribution is, what their population density is, and seasonal migrations of the animals in the region. We should understand much of the basic ecology of the area, the interrelationships of all the animals with their environment; the structure of biotic communities; the dynamics of population fluctuations; the food chain—energy flow and factors; and much more. . . . We should take the opportunity to study the biology of the region before the detonation, so that we can properly evaluate the after-effects of the blast, from both the physical and radiation aspects as they affect the regional ecology.[38]

The University of Alaska proposal thus largely conformed to the notion of preshot background studies that had been suggested by Johnson, even as it took a more sophisticated view of exactly what constituted "background" for the proposed experiment in ecology. The University of Alaska proposal requested $107,000 (for the first year alone) for biological and ecological field studies in the Cape Thompson region. But although the AEC may have been a major funding agency for ecological research at this time, it was not in the habit of giving away quite so *much* money for ecological surveys. For instance, prior to the installation of reactors for fuel production at the Savannah River Plant, in 1951 Eugene Odum had proposed a far-reaching study of the ecological effects of low-level radiation that was budgeted at $267,000 annually, only to end up with a far more modest annual grant of $10,000 for more rudimentary inventories of animal populations. But the Fairbanks biologists had some leverage, and they took advantage of the opportunity to fund their research. With DBM ecologist Wolfe's support, the proposal was approved within two months.

So again Project Chariot was transformed in response to resistance. When Livermore's Plowshare teams first went to Alaska in July 1958, resistance had come in the form of skepticism over the project's dubious economic rationale, in the very absence of roads and railroads for the resource extraction periphery that Livermore expected their nuclear harbor project to serve in northwest Alaska. But Livermore and the AEC responded by scaling down the project into a smaller, "early and obvious demonstration" of the safety and utility of Plowshare techniques. The development of the Chariot Environmental Program was a response to different kinds of resistance from both inside and outside the AEC. For many of the Livermore physicists, chemists, and engineers working on Plowshare, there may have been little interest in Cape Thompson's local environment besides its providing a place to explode large bombs.[39] Yet a program for pre- and postshot ecological surveys, insiders at the DBM had argued, would be useful to the aim of developing a practical nuclear excavation technology and, at the same time, to sponsoring the development of biological and ecological knowledge. The University of Alaska biologists had presented Livermore with a more oppositional form of resistance, but this, Plowshare scientists like Johnson hoped, could be overcome by including them in the project, to a degree, through lucrative research contracts in their areas of specialty. In turn, shortly after the scientists were enrolled in Chariot, they could even be listed as local experts taking part in the studies to ensure Alaska's future health and welfare.[40]

The Fairbanks biologists were not unwitting dupes in this process. Bill Pruitt, a mammalogist in the environmental program and an eventual outspoken opponent of Project Chariot, later recalled, "There are a number of people who, over a period of quite a number of years, had been trying to do Arctic research—field biology—living from hand to mouth, trying to pick up grants here and there. So that when this rather magnificent grant, or contract, came through, of course they jumped at it. Yeah, I did too. Sure. You know, a good field biologist will do damn near anything to get back in the field again."[41] And yet given the confrontational stance already adopted in the meetings with

Keller and Shelton, it is possible that, for some of the Fairbanks biologists, a rigorous environmental program was seen both as a means of funding their own costly Arctic fieldwork *and* as a potential means of subverting Chariot. By forcing the AEC to make a careful, scientific assessment of the biological costs of the proposed blasts, some of the biologists may have believed it possible to discover for themselves whether the project was as safe as Livermore had claimed and, if it was not, to better understand and publicize the risks and uncertainties involved. For Livermore, the problem of controlling the kind of involvement in Chariot of the Fairbanks scientists would depend on the establishment and maintenance of circumscribed rights and rules of authority to evaluate the information produced in the environmental program.

THE COMMITTEE ON ENVIRONMENTAL STUDIES

The studies outlined by the Fairbanks biologists were incorporated into the AEC's plans for the Chariot environmental program during the first months of 1959. The program would eventually include more than forty separate studies, with researchers drawn from the University of Washington, the University of California, Los Angeles, the U.S. Fish and Wildlife Service, the U.S. Public Health Service, and the Office of Naval Research (in addition to the University of Alaska and the DBM), encompassing an impressive breadth of interdisciplinary, regional expertise in biology and ecology, radiology, marine studies, meteorology, geology, and geography. The substantive purposes of the environmental program (including but not limited to the University of Alaska studies) were

> A. To determine the optimum time of year for the Chariot detonation and the most desirable orientation of the fallout zone. These determinations will be predicated on causing the least amount of damage to the populations of animals, birds, fish and plants in the affected area and minimizing subsequent perturbations in the present balance of the ecosystems. Special attention will be give the habits and requirements of the local human population.
>
> B. To determine, insofar as possible, the actual biologic effects of a nuclear detonation on an area by contrasting post-shot ecology with pre-shot ecology. This knowledge, safely obtained in a remote area, is vital to the extension of nuclear excavations to areas of greater habitation.[42]

Since concerns for the "local human population" were crucial to the AEC's assurances of public safety, a human geographer was added to the program whose work would examine "the relationship between human economy and the biological environment" in the Inupiat villages of Point Hope and Kivalina.[43]

Pivotal to the organization of the research, all of the scientists in the environmental program were to have the results of their work channeled through an AEC-appointed Committee on Environmental Studies for Project Chariot. The committee was to be chaired by the DBM ecologist John Wolfe, and included five ostensibly independent experts in radiology, radiation biology, ichthyology, geology, and public health, all of whom, with the exception of Wolfe (as well as committee secretary Ernest Campbell), had research experience in polar regions.[44] Its main task was to plan "the scope of bio-environmental studies in support of Project Chariot, [and] to obtain sufficient information for prediction of the effect of the project on the biosphere, both immediately following the detonation and for the future." As part of its duties, the committee would also receive and review reports from all principal investigators and, in turn, "summarize these findings to the Manager, SAN [AEC San Francisco Operations Office], together with the committee's recommendation for further action, if any." The committee was thus in a position to recommend which studies were essential to the program and which were not, and to evaluate and assimilate the results of those studies. AEC program managers and public information specialists would then "authorize public dissemination of the committee's consolidated report to interested public officials, the scientific community, and news media."[45]

An institution of some kind—and perhaps ideally, an interdisciplinary committee—was clearly needed to organize and oversee such an extensive and diverse environmental program. Just as important, the committee was also, at least potentially, a means of controlling aspects of the studies, its power residing in the privilege of deciding which scientific results should be brought to bear on the evaluation of Chariot.

Before any animal populations were catalogued, vegetation plotted, or Inupiat villages visited, the AEC's public relations machine intended to take full advantage of the idea of the Chariot Environmental Program. In a March 6, 1959, press release, the AEC announced the formation of the environmental program, and in a June 24 statement, it explained in greater detail the studies that were to precede Chariot, listing the impressive scientific agencies and personnel who would be involved in the preshot biological, geological, and engineering studies. "Project Chariot," as it now stood before the public, one year after Teller and his colleagues' first tour of Alaska, was "an experiment to determine the factors involved in the use of nuclear explosives for excavation purposes as well as [to] demonstrate the excavation of an area that could be used as a harbor."[46]

Surveys at the Chariot site at Ogotoruk Creek, to be carried out in 1959 and possibly 1960, were expected to serve four major objectives:

1. Obtain engineering and geological data.
2. Make a biological inventory of plant, animal, and bird life of the sea, land, and fresh water in the site vicinity.
3. Identify the food chains and other ecological features of the region.
4. Study human geography and habit.

"If the . . . studies prove the safety and practicability of the experiment," continued the press release, "it could be conducted in 1961."[47]

The environmental program was somewhat ambiguously located within a larger phase of preshot research that included technical and engineering studies, logistical work, and a public safety program.[48] The results of contract science carried out under the environmental program were to be funneled through Wolfe's committee and, in turn, this work was to be evaluated, along with the other preshot programs, by coordinators at Livermore and by the AEC's testing specialists in Nevada and elsewhere, all managed under the organizational umbrella of the AEC's San Francisco office.[49] But if this structural experimental space was to *work*, it could not occur on paper alone; rather, the point was precisely to attach a system of authority and evaluation to the control of the physical, biological, and social environment at Cape Thompson.

Second Plowshare Symposium

The weapons scientists at Livermore did not want to badly contaminate the environment around Ogotoruk Creek. Apart from any ethical misgivings they may or may not have had about injuring the Alaskan Arctic and its inhabitants, to do so would only have served to create more resistance to the nuclear earthmoving program, when the rhetorical thrust of Chariot, by demonstrating just how safe (not to mention spectacular) geographical engineering could be, was to remove potential resistance to larger projects elsewhere. But while it is important to recognize that project scientists, in their ambitious efforts to turn the bomb into something good, did not wish to do harm, it is also important to understand the experimental and *uncritical* nature of their approach to Chariot's radioactivity hazards. As such it is all the more important to examine how Livermore and other AEC scientists understood (and constructed) the problems posed by Project Chariot, from the challenge of nuclear cratering itself to that of evaluating and controlling radioactivity hazards, in ways that allowed them to insist that the instant-harbor project, and nuclear excavation in general, could be conducted safely.

While biologists in Fairbanks still awaited formal approval of their proposal for field studies at Cape Thompson, much of the rationale for Chariot and the entire Plowshare program was extolled at the Second Plowshare Symposium, a conference put on jointly by Livermore (now called the Lawrence Radiation Laboratory at Livermore) and the AEC San Francisco Operations Office on May 13–15, 1959, which was held in both Livermore and San Francisco.[50] The symposium provided an opportunity for Livermore scientists and others in the AEC's classified community to convene and report on progress to date in their paper studies, and to discuss ideas, plans, and projects for the future. Only the first day of the conference, on the topic of underground nuclear explosions, seems to have been classified; the next two days, which focused on "putting this knowledge to work" in nuclear excavation, power recovery from nuclear explosions, mining applications and petroleum recovery from oil shale, were public scientific meetings complete with published proceedings.[51] With the nuclear test moratorium in its seventh month, perhaps the AEC wanted to clarify that, contra the skepticism encountered the previous year at Geneva, it really *did* have an active research program in peaceful nuclear explosives that could be exhibited separately from the weapons program. Indeed, the hundreds of pages of scientific papers that constituted the unclassified portion of the proceedings give a fair indication of the extent that Plowshare ideas had been developed and codified by this point, but they also suggest a certain ambiguity regarding how biological and ecological questions concerning fallout from nuclear excavation were to be resolved.

NUCLEAR EXCAVATION AND PROJECT CHARIOT

Putting the knowledge of nuclear explosions to work began with a broad assessment of progress in nuclear excavation, with studies presented mainly by Livermore and Sandia Corporation scientists and engineers that included "Excavation with Nuclear Explosives," "Long Distance Blast Prediction Problems," "Cratering with Chemical Explosives," "Excavation of a Sea-Level Ship Canal," and "Excavation of a Harbor (Project Chariot)," along with three studies that were concerned chiefly with the problems of fallout and the distribution of radioactivity.[52] Together, they amounted to something like the state of the art in nuclear earthmoving at a moment of profound optimism—to borrow from the vocabulary of Edward Teller—over the Plowshare program, even though no earth had yet been moved with "peaceful" nuclear explosions.

Just what did participating scientists know about nuclear excavation? High-yield nuclear tests in the Pacific had produced large underwater craters (often the remains of annihilated atolls), but little had been learned about crater-formation processes from these events.[53] The development of the first general laws of nuclear cratering was based instead on four low-yield explosions, none larger than 1.2 kilotons, conducted at varying shallow depths in the desert soil of the Nevada Test Site from 1951 to 1958.[54] From these experiments, and from cratering tests conducted with chemical high explosives, preliminary "scaling laws" were developed to predict variations in crater width and depth as functions of nuclear explosive yield, depth of burial, and soil medium. "It is generally concluded," wrote Livermore's Melvin Harrison, "that crater diameter is proportional to the cube root of energy release and crater depth proportional to the cube root or fourth root of the energy."[55] Such scaling laws could be used to provide "optimum depths of burial" for explosives of variable yield in relation to the desired shape of the crater, depending also on variations in soil characteristics and in the makeup of the bombs themselves. But this formula for the optimum placement of explosives also had to be developed in relation to safety considerations concerning fallout and blast effects (such as seismic and air blast hazards). Fallout was the most serious potential problem to do with nuclear excavation, but Livermore scientists believed it could be adequately contained through two primary techniques involving, on the one hand, reduction in the release of radioactive materials, and on the other, strategies for emplacement of the explosives that would minimize the extent and distribution of radioactive fallout. The first means of reducing radiation hazards was in the design of cleaner explosives, the ongoing (classified) Livermore project that included reductions in fission-to-fusion ratios as well as techniques for wrapping the explosive in a neutron-absorbing substance such as boron. But fallout could also be contained, it was argued, through a better understanding of the dynamics of shallowly buried nuclear explosions. The fundamental principle was that the deeper the explosive was buried, the less radioactive materials would be vented into the atmosphere. The optimum depth of burial for crater size could be adjusted, it was argued, at some expense to the total volume of earth moved, to contain more "fissile materials" underground while still producing a "respectable crater size."[56] In addition, Livermore scientists expected that the largest radioactive particles blasted out would tend to "collapse and fallback" into the crater and in its near vicinity (see figure 2.3), thus limiting the diffusion of radioactivity along with the containment and reduction of the total release of radioisotopes.

Still, it was "obviously a wide stretch of one's imagination," as Livermore's Harrison commented, "to predict a great deal about the variation of crater width and depth as a function of nuclear yield and depth of burial" on the basis of just four relatively small nuclear tests.[57] And although some useful data could be extrapolated from experiments with high explosives, this said nothing about the dynamics of radioactivity in earthmoving blasts, nor about the qualitative differences between conventional and large-scale nuclear explosions. So the development of scaling laws and emplacement techniques was of

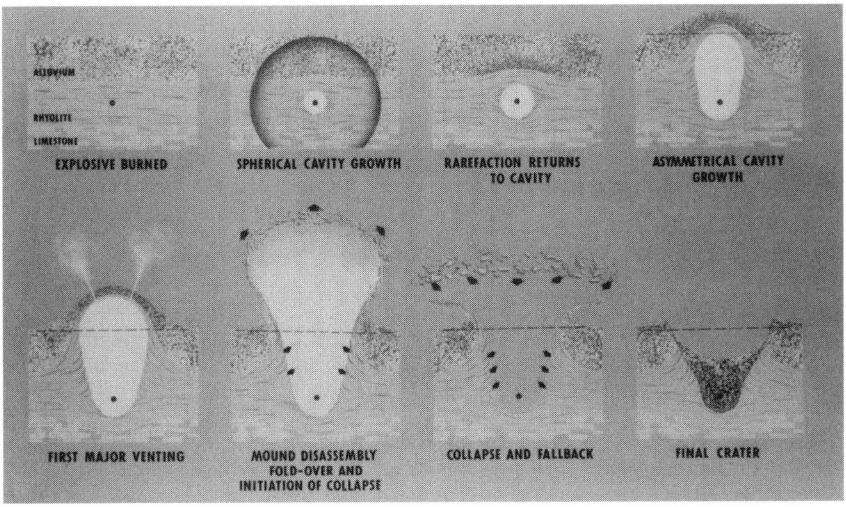

Figure 2.3. Crater formation history. Livermore scientists' phenomenology of a nuclear cratering explosion. *Credit must be given to the University of California, Lawrence Livermore National Laboratory, and the Department of Energy (Atomic Energy Commission) under whose auspices the work was performed when this information or a reproduction of it is used.*

limited value without the opportunity to test the ideas on a large scale and to compare observed experimental data with predictions (for example, of crater dimensions; particle size and distribution of fallout; or the mechanics of the crater's slope stability) through nuclear experiments designed explicitly for Plowshare. What was already clear, by this point, was that nuclear excavation would only be economically viable for massive earthmoving jobs. The economic cost-per-energy unit of nuclear explosives decreased logarithmically so that only after the size of explosive yield was in the hundreds-of-kiloton range could it undercut the cost of conventional high-explosive techniques, with substantial savings coming only via blasts in the megaton range.[58] The original megaton-scale harbor blast for Chariot, with its pretense of economic viability, had been designed on this basis. But since, as Livermore's Clifford Bacigalupi understood it, the "major objective of the experiment is to obtain cratering and effects information that can be applied to future excavation work," the size of the blasts had been reduced. The issue of Chariot's economic viability was set aside and was scarcely mentioned in symposium proceedings.[59]

Although some of this information to be produced by Chariot was needed for determining scaling factors for determining optimum depths of burial for future projects and experiments, much of the "effects" information that Plowshare scientists wished to obtain concerned the movement of radioactivity from an earthmoving blast, a problem that, by their own standards, they knew little about. Vay Shelton, for example, cautioned the symposium audience (and *Proceedings* readers) about the curve he had graphed to relate "scaled depth of burial" to radioactivity release by specifying that "the shape of the curve between 70 and 230 feet is theoretical and should be used with care. We will find out more about this as time goes on, particularly at Chariot."[60] And yet despite the explicitly experimental nature of this evaluation of radioactivity hazards "at Chariot," there seemed to be little doubt among Livermore's Plowshare scientists that the fallout problem was one that could be solved, in a technical sense, and that nuclear excavation would eventually be freed from the spatial fetters of remoteness. As Harrison saw it: "There are many possible nuclear excavation sites available, some quite near civilization. For the moment we prefer to conduct experiments in isolated sites, though with experience we are certain we can operate relatively close to population centers."[61]

But the Chariot site was not at all isolated from the village of Point Hope, just thirty-one miles from ground zero, a fact that did not seem to register under the category of "population" for some of the symposium scientists. Bacigalupi described site-selection procedures for Chariot:

> Prior to going into the field to select the actual location for the excavation, it was necessary to select the general area. In this selection consideration was given to the remoteness of the site from *any existing population* and that the land be government-owned.
>
> Using these criteria as a general guide, it was determined that the northwest coast of Alaska between Nome and Point Barrow met the above requirements. A geological consulting organization was then asked to investigate this region to determine the location that would provide the best site. As a result of this survey, the area between Cape Thompson and Cape Seppings was selected. Within this portion of the Alaskan coastline three sites were tentatively selected from existing contour maps and aerial photographs. During the summer of 1958 each of the three sites was visited and it was determined that, from an engineering standpoint, Ogotoruk Creek provided the best terrain for the experiment.[62]

With the vestigial economics now removed from the project, the "engineering standpoint," it seemed, together with remoteness, had always been the central factors in the selection of Ogotoruk Creek. Inupiat people living in the region failed to register as an "existing population"; perhaps they had been easy to miss in the process of scanning contour maps and aerial photographs. Instead, the importance of the engineering standpoint in site selection comes to the fore, with no mention of mineral deposits or safe havens. At the Second Plowshare Symposium, Chariot appeared to have only ever been an experiment in nuclear harbor excavation.

The creation of the model harbor at Ogotoruk Creek was a test based on partial knowledge of nuclear cratering processes, extrapolations and predictions based on what was known already, that would, in turn, provide new data to confirm, deny, or otherwise clarify hypotheses for establishing scaled depth of burial, better understanding the mechanisms of fallout distribution, and for testing various other predictions. Many of the laws developed in these paper studies would be tested for the first time with Chariot, or at least for the first time at a scale in the hundreds-of-kiloton range: the use of explosives in the 100- or 200-kiloton range would elevate the scale of the energy release over the prior craters in Nevada by two orders of magnitude. As it stood, the experiment called for two 200-kiloton explosives to be buried at an actual depth of 700 feet (or a *scaled* depth of 120 feet), spaced 775 feet between the centers, to produce the "turning basin," with a row of four, 20-kiloton explosives buried 375 feet deep and 408 feet apart to provide the channel into the model harbor. This particular spacing of the explosives was designed to shape the harbor floor (albeit not in the form of a polar bear) by minimizing the cusps formed by overlapping circular craters.

Theoretically, the Chariot explosions were expected to vent above ground anywhere between 5 and 20 percent of the total radioactivity produced, depending on how deeply the explosives were buried. The problem was that while scientists believed, with reasonable certainty, that this fallout could be directed inland and away from nearby Inupiat villages, they conceded that outside two hundred miles their predictions had little meaning. Breaking the moratorium for a peaceful nuclear explosion was one thing, but to do so and at once send large amounts of fallout over Siberia, in the event of a reversal of wind direction, was quite another. Perhaps for this reason, a "deep model" for depth of burial was favored, which could be expected to intensify fallout within fifty miles but reduce it outside that distance. The preliminary "fallout sector" constructed for Chariot (figure 2.4) reflected this deep model with an estimated 17 percent of the bombs' radiation vented.[63]

The fallout map also reflected a particular geography, a construction of space wherein the earth's surface is remapped as a repository for fallout and not much else. Of course, mapping is always a process of selection, of representing spatial distributions of some landscape features and excluding others.[64] On this map, a temporal dimension is portrayed as well as the spatial by illustrating contours of estimated gamma-ray doses at selected time intervals, each with an associated level of fallout attached to an elliptical spatial zone. And yet the rest of the landscape—with its human population reflected only in parenthetical

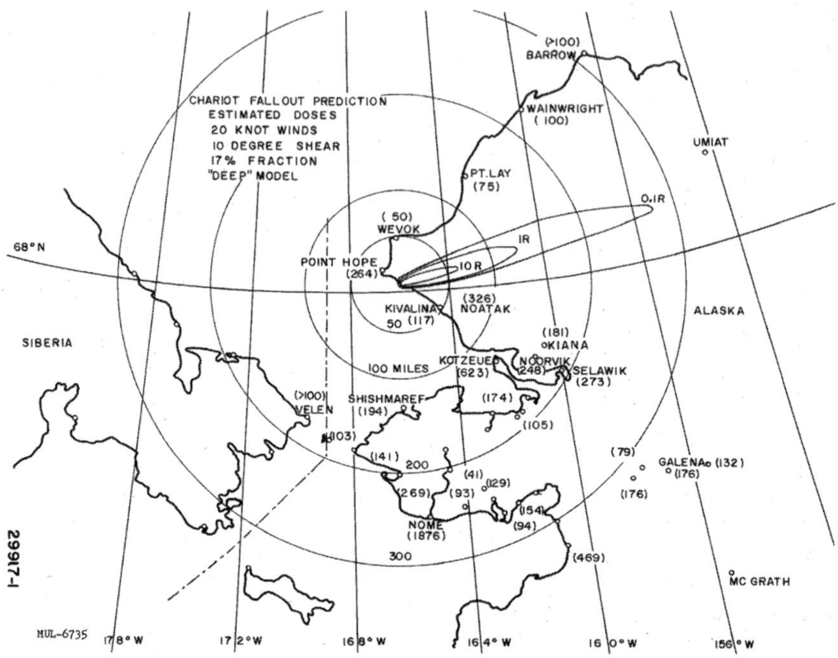

Figure 2.4. Predicted fallout sector for Project Chariot. *Source: U.S. AEC,* Proceedings of the Second Plowshare Symposium, May 13–15, 1959. *DOE doc. NV0015962, Nuclear Testing Archive, Las Vegas, Nevada.*

numbers alongside the names of towns and villages (all seemingly distant from the deposition of fallout) and its animal and plant populations not portrayed at all—appears static. The technology of mapping fallout sectors, a necessary component of radiation safety practices for nuclear testing, thus carried with it particular, anti-ecological conceptions of space that contributed to whatever "sense of place" the Livermore scientists were developing of the Cape Thompson region, even as the primary task of the maps remained largely pragmatic, allowing those seeking technical solutions to the fallout problem to more or less get on with their work.

As Bacigalupi reported, the AEC would review the "geological, biological, and engineering information" obtained from the Chariot site during the 1959 season and, if all went well, "construction for the device [explosive] emplacement can be scheduled to take place during the summer of 1960. The actual detonation could follow in the spring of 1961. The majority of the proposed detonation analyses could be completed in the summer of 1961."[65] Detonation in the spring would allow the AEC to complete its postshot analyses before the Alaskan days became too short when the Arctic fall and winter arrived. But what if the "information" brought back from Cape Thompson suggested a geography much different from that of the fallout sector map, one emphasizing the ecological dynamism of the landscape and illustrating a balance of interconnections between places, and between species, that might be upset by the blasting of the model harbor? This question was not raised by Bacigalupi or any of his Livermore and Sandia colleagues, but did appear in John Wolfe's symposium paper on "The Ecological Aspects of Project Chariot."

ECOLOGY LESSON

"The integration of bio-environmental knowledge into projects that Dr. Teller has described as 'geographical engineering' has not always been accomplished. But it is inconceivable," Wolfe declared to his colleagues, drawn mainly from the physical and engineering sciences, "that undertakings of this technical magnitude should ignore the living aspects of geography, and for this reason the Atomic Energy Commission has a program for taking these matters into account."[66] It is difficult to pin down just what he meant in his reference to geographical engineering projects of the past that had failed to take bio-environmental knowledge into account. Was he referring to harbors and canals excavated by conventional means, or to other subterranean nuclear tests? Whatever the case, in stressing the need to account for the "living aspects of geography" affected by massive environmental transformations, Wolfe was presenting a construction of space very different from that represented by the proposed Chariot fallout sector. Fresh from a March visit to Fairbanks during which he had conferred with University of Alaska biologists, among others, on the development of the environmental program, Wolfe's talk at the symposium began as a lesson in Arctic ecology. Against conventional notions of the Arctic tundra as remote, desolate, and barren, he emphasized to his colleagues:

> It is not *remote* to the Eskimo, the arctic fox, the ptarmigan; the flowering plants blooming by the thousands per acre in brilliant colors during the early period of wet earth belie its *barren*-ness and *desolation*, and represent the showiest natural floral display of any vegetation in the world; and the tundra climate is most salubrious to indigenous dwellers among which are some of the majestic animals of the earth, not excluding man.
>
> The tundra, and its adjacent waters, is an ecological system. Landscapes are dominated by various communities of plants, all small or dwarfed by genetic and/or environmental conditioning. Predominant are sedges, grasses, heathers, mosses and lichens. These are the primary food producers of terrestrial habitats. The primary producers of the sea are algae, and it is upon algae that the entire human economy in the arctic rests.

Wolfe's diagram of the Arctic ecological system illustrated the basic interconnectedness of the biological environment, with the Eskimo at the top of the food chain. "It is in this environmental complex," he noted, situating the matter, "that it is proposed to conduct certain engineering experiments relative to excavation of large quantities of the earth's crust by means of nuclear devices."[67]

What was the point of Wolfe's ecology lesson? What role did he envision for bioenvironmental knowledge in this geographical engineering project? Answers to these questions came across rather ambiguously in Wolfe's paper, perhaps because the ecologist was careful not to overstep his bounds: "There is ample evidence that such undertakings are technically feasible. Biologists are not competent to advise on the why or how of such projects from the engineering viewpoint. But they do raise the question of biological cost and can provide data as to when, where, and under what biological conditions such an effort could best be accomplished from a total environment point of view." Throughout his paper, Wolfe spoke of the need to protect not just Eskimos but also the biological environment that was their resource base. He evoked Aldo Leopold's notion of the land organism (slipping at times, however, into discussion of a land *mechanism* instead), and he warned, with an awareness rare in the symposium's *Proceedings*, that "it is scarcely too early in the Atomic Age to give considerable attention to the environment which supports man on this planet. Phylogenetic lines have terminated in the past; and will continue to do so for many of the same reasons."[68]

With the ecological stage thus set, Wolfe described the interdisciplinary environmental program that he was rapidly organizing at the time. The program was divided into five categories: biological inventory of the Cape Thompson area; mapping and photography (including the plotting of vegetation patterns and mapping of animal concentrations and Eskimo hunting grounds); "dynamic aspects of the biological assemblages" (including investigations of food webs and other ecological relations); studies of human populations; and studies of the physical environment. As a perusal of the forty-one studies that were eventually carried out under the Chariot environmental program, listed in table 2.1, makes clear, Wolfe was hardly exaggerating when he boasted that the "program probably represents the largest single integrated effort ever made to study environment, physical and biotic, in an area of this size." Wolfe's sensitivity to the unique ecology of the region, however, did not make the business of modifying that environment with nuclear explosives any less attractive. Rather, the value of the experiment was "further enhanced," he continued, "by the fact that [it] is being conducted in an essentially primeval area, that is, the living population—their fluctuations, periodicities, densities—are at essential equilibrium with the arctic environment." For environmental scien-

Table 2.1
Project Chariot Environmental Program, 1959–1962: Scientific Studies Completed by Category

Area of study	*Physical characteristics*	*Biotic systems and bioenvironment*
The Land	Weather and Climate Geographic Setting Areal Geology Engineering Geology of the Chariot Site Drilling, Coring, and Frozen-core Analysis Groundwater Hydrology Surface-water Discharge of Ogotoruk Creek Chemical Character and Sedimentation of the Waters Permafrost and the Geothermal Regimes Seismic, Gravity, and Magnetic Measurements Slope Movement and Related Phenomena Soils of Ogotoruk Creek Watershed	Vegetation and Flora Palynology of the Ogotoruk Creek Area Freshwater Algae and Zooplankton Limnology of Tundra Ponds and Ogotoruk Creek Avifaunal Investigations The Caribou of Northwestern Alaska Ecology of Terrestrial Mammals Terrestrial Invertebrates
The Coast	Arctic Beach Sedimentation	Sea-cliff Birds Zooplankton of Some Arctic Coastal Lagoons
The Chukchi Sea	Oceanography of the Southeastern Chukchi Sea Geology of the Southeastern Chukchi Sea Marine Geology of the Ogotoruk Creek Area	Net Plankton Volumes in the Chukchi Sea Benthic Invertebrates of the Southeastern Chukchi Sea The Ascidians Fishery Investigations of the Southeastern Chukchi Sea The Salmon Resource Marine Mammals
	Physical and biological characteristics	
The People	Archaeology of Ogotoruk Creek Human Ecological Investigations at Kivalina A Human Geographical Study Meat Consumption at Three Northern Eskimo Villages	
Radioactivity	Aeroradioactivity Survey of the Lisburne Peninsula and Adjacent Areas Adsorption Equilibria between Earth Materials and Radionuclides Radiological Analysis of Marine Organisms Radioactivity in Northern Alaskan Eskimos and Their Foods, Summer 1962 Radionuclides in Terrestrial and Freshwater Biota	

SOURCE: Categories and titles of studies as appearing in N. J. Wilimovsky, ed., and J. N. Wolf, asst. ed., *Environment of the Cape Thompson Region, Alaska* (PNE-481, Division of Technical Information, United States Atomic Energy Commission, 1966).

tists trying to understand relations, change, and process, the principal benefits would accrue *after* the Chariot nuclear blast—from postshot studies, whatever the damage done. "If radioactive fallout should become a major factor, a rather complete knowledge will be required of other factors, disturbances, and natural trends in geological, meteorological, marine, and pedological processes."[69]

There is no evidence to tell us how Wolfe's paper was received at the second symposium. Whatever qualms the participants may have had about Wolfe's "total environment point of view," some may still have been stirred by his concluding remarks, in which he quoted at length from President Thomas Jefferson's instructions to Captain Meriwether Lewis regarding the need for environmental observation and measurement during the Lewis and Clark expedition west. If the joy of geographical engineering, for the Livermore physicists, could be found—as Teller would tell an audience in Fairbanks just a few days later—in efforts to "reshape the earth to your pleasure," then for Wolfe, the curiosity and the pleasures of the environmental program were to be found in tracking the changes wrought on the ecosystem by this reshaping.[70] But what did raising the question of "biological costs" mean if, at the same time, bioenvironmental scientists were simply expected to provide data useful for deciding when, where, and under what conditions—but not why or if—the experiment should be conducted? The seeming contradictions of ecology, untroubling to Wolfe, would not be as easily resolved for all of the environmental program's contract scientists, some of whom would come to see the disparate constructions of space in the Chariot "preshot studies," that is, the fallout sector and the "living geography" of the ecosystem, as fundamentally incompatible.

On the Coattails of Nuclear Excavation?

In early July 1959 the AEC contractor Holmes and Narver began construction of a base camp on Ogotoruk Creek for Chariot's preshot studies: facilities would ultimately include sleeping quarters for eighty-four in Jamesway huts, a mess hall, radio communications facilities, a small biology laboratory, an airstrip, and other support facilities (see figure 2.5).[71] Although the biology field lab would be useful for some researchers, such as those making scat and stomach-content analyses for the purpose of piecing together food webs, we could also say that the entire Ogotoruk Creek–Cape Thompson region was being turned into a kind of laboratory as researchers in the environmental program began their work in the summer of 1959. University of Alaska biologists and their field assistants had actually arrived three weeks before construction

Figure 2.5. Base camp for Project Chariot environmental studies on Ogotoruk Creek, Cape Thompson, Alaska (viewed from above). *Credit must be given to the University of California, Lawrence Livermore National Laboratory, and the Department of Energy (Atomic Energy Commission) under whose auspices the work was performed when this information or a reproduction of it is used.*

of the base camp started, and were living in tents as they initiated their fieldwork. While geologists and hydrologists engaged in the difficult work of digging core samples from the permafrost soil under Ogotoruk Creek, the biologists initiated an intensive mapping of the flora and fauna of the region: setting up numerous, representative one-acre plots for inventorying all manner of species, hiking transect routes for counting birds and bird species, tracking caribou herds (from the air, whenever possible), acquiring samples of animal pelts and skeletons, and many other field practices. If Wolfe's paper at the symposium had underscored the interconnectedness of life in the Arctic, then the first season of fieldwork at the Chariot site was beginning to piece together more precisely the nature of these ecological relations.

Before the AEC funded a *second* season of this rather expensive Arctic fieldwork, it wanted a preliminary report from Wolfe's committee, to suggest "whether the project should proceed or be indefinitely suspended."[72] The AEC needed to know, or at least needed the committee's assurance, that Chariot was still feasible in terms of the problem of "biological costs." On January 7, 1960, the committee released a public statement:

> It is the unanimous opinion of the Committee, based on data known to it as of December 10, 1959, that Project Chariot may be carried out under the following conditions:
>
> 1. Time—The preferred time of the year for the detonation is Spring, i.e., March or April.
> 2. Debris—The preferred disposition of the debris, especially that of a radioactive nature, is to sea; placing the debris to land is also considered acceptable.
> 3. Qualifications—In no case should radiation be delivered to humans in excess of that specified as acceptable for the general public; nor should the detonation cause significant damage to the food sources of the indigenous human population.[73]

After explaining that these considerations were based on the objective of minimizing biological costs, the recommendation for a spring detonation was justified for six reasons:

> 1. Few birds in the area.
> 2. Most small animals under snow cover.
> 3. Most plants are under snow cover and their metabolism is low.
> 4. Local hunting activity on the land is low.
> 5. The sea and inland waters are under ice; snow is on the ground. It is expected that radioactive debris will be flushed from the frozen landscape by the spring run-off of rapidly melting snow.
> 6. Weather is generally good and daylight is increasing which will facilitate project studies.

Finally, the statement added the disclaimer that these recommendations were based on incomplete data, and that continuing environmental studies would be required to "provide the basis for firm recommendations." Uninterrupted preshot studies until March 1961, the committee said, would be crucial to the "development of adequate baselines from which post-shot evaluations may be made."[74]

Even taking the call for continuing studies into account, many of the biologists working under the University of Alaska contract, including botanists Albert Johnson and Les Viereck, mammalogist Bill Pruitt, and ornithologist Gerry Swartz, as well as human geographer Don Foote, an independent AEC contract scientist, were shocked by the committee's statement, and angry over what seemed to be not only a misinterpretation of their preliminary reports but also a violation of the ethical and procedural standards that governed scientific practice. Indeed, the qualification that data were incomplete was perhaps the only aspect of the committee's statement that they did agree with.

In the first place, given the early stage of the bioenvironmental studies, it seemed ill-advised to set forth conditions under which the blast could be safely conducted. More troubling still were the justifications for a spring detonation, which were not supported by present data—not least because no environmental studies had yet been conducted during these months—and which in some ways contradicted historical data. Reasons 1 through 5 in the list of justifications would subsequently be shown to be almost entirely wrong: birds would be observed in the area numbering from 1,000 to 2,000 per hour; squirrels would be seen as early as March; because of seasonal high winds, snow cover was *not* necessarily high at this time; and consequently lack of snow cover in certain areas made them especially attractive places for caribou grazing and, in turn, for Inupiat hunting (an activity that was not low at the time, but actually quite high).[75] The only reasonably accurate justification was the sixth—that increasing daylight hours would facilitate postshot project studies—and that statement, as Livermore's Bacigalupi had suggested at the second symposium, was among the reasons that the AEC had favored a spring detonation in the first place.[76]

Why Wolfe's committee would advance such a statement at this time is difficult to say. Had it been put to the committee, by Livermore scientists or AEC technocrats, that no aspect of Chariot could move forward without some assurance of safety at this early stage? Or was it that, for Wolfe and some of his colleagues on the committee, what really mattered was that the preshot "baseline" studies continue uninterrupted, by whatever means, for the benefit of postshot ecological investigations? In a letter to University of Alaska zoologist Brina Kessel, Wolfe later shed some light on his understanding of the environmental program in its relation to the larger aims of Project Chariot: "There would not have been any [environmental] program such as this at all if it were not for the *idea* of nuclear excavation. We still ride on the coat-tails of that idea, since the actual detonation has not received approval. We are obligated therefore, to support researches useful to the idea but at the same time we have insisted on studies contributory to science."[77] Wolfe thus appealed to the ideal of (disinterested) science as a way out of the problem of interested (or obligated) research.

Whatever the course of events that led to the committee's January 7, 1960, statement, it created a new environment of mistrust among many of the contract scientists. For the botanist Viereck, who had been recruited from his doctoral research at the University of Colorado by the Fairbanks biologists to take part in the studies, the January 7 preliminary endorsement of the spring detonation "was the first indication that any of us had that the Environmental Committee was merely going along with the predetermined policy of the

AEC and was not using the reports of the investigators to formulate their decisions."[78] Perhaps naïve before the committee's statement, and seemingly removed from mainstream debates over fallout during the late 1950s, after January 7 Viereck and several other young scientists became increasingly uncomfortable with the degree to which *their* science was being made "useful to the idea" of nuclear excavation. In the next chapter, we will examine how these tensions were resolved through politically and scientifically oriented oppositional movements; for now, the point is to stress how the terrain of debate was developed through the complex interplay of heterogeneous actors, institutions, and scientific perspectives, and to note that this "terrain" of debate was not entirely a metaphoric one. The problem, in a sense, was this: there were two disparate scientific spaces present, the physical space of nuclear excavation, in which the Cape Thompson region was understood chiefly as the object of a geographical engineering experiment; and the biological space of regional ecology, in which living interrelationships in nature and space were stressed. But there was only one place for these contrasting scientific visions to reside in and reconstruct: the region starting at the mouth of Ogotoruk Creek and extending as far as the debris and fallout produced by the instant-harbor blasts would carry.

A Demonstration Transformed

One environmental historian has suggested that the Project Chariot bioenvironmental program amounted to something like the first environmental impact statement.[79] A decade before the passage of the National Environmental Policy Act, the breadth of interdisciplinary environmental studies developed to evaluate the potential effects of Chariot on land and marine ecosystems stands out as a model for understanding the far-reaching consequences of technological transformations of environments. But it did not start out this way for Project Chariot. In fact, it was precisely through resistance to Plowshare program plans—and through Livermore's strategies for *overcoming* resistances—that the environmental program took shape as it did.

The Plowshare program moved to northwest Alaska for a number of reasons, all of which were designed to satisfy the interrelated (perceived) needs of experiment and demonstration, science and spectacle, in nuclear excavation. But for Chariot's planners at Livermore, it quickly became clear that this "early and obvious demonstration," as Teller had put it, would take time to develop, far more than the original eighteen-month plan for demonstrating a "useful excavation" would allow.[80] It was also clear that Livermore would not be able to achieve the project with AEC support alone. As Plowshare moved to Alaska, it encountered resistances, first economic, and then scientific (and explicitly biological), to what, among the weapons designers at Livermore, had seemed like a capital idea: the geographical engineering of a working harbor at Cape Thompson. And so the justification for the project shifted and, along with it, the nature of the experiment itself. Still, Livermore's ability to carry out Chariot depended on establishing the claim that the project would be executed safely "with respect to persons within and outside the immediate area and to fish and wildlife."[81] Not only was the scale of the model harbor reduced to a much smaller hole in the ground than originally planned, but wide-ranging ecological surveys were integrated into the design of the experiment as well. But in this regard, the environmental studies were set up with clearly delineated procedures and rules of authority geared, in large part, to co-opting the Fairbanks scientists' concerns for biological costs into the AEC's related, but not identical, concern for useful biological information.

After the January 7 statement, controversies emerged, on the one hand, over whose expertise counted concerning the potential biological effects of nuclear explosions at Ogotoruk Creek and, on the other, over the political, legal, and procedural boundaries set in place to govern the processes of fact making and of dissent in connection with the environmental program. Beyond the power to demarcate this experimental space in Alaska, the power to actually see Chariot through to completion hinged on the questions of who could speak authoritatively for both the scientific spaces and the geographical place of the Cape Thompson region, and from whom legitimate dissent was possible. Here the geographies of controversy would coalesce into a contested politics of place.

Chapter 3 Geographies of Authority

Livermore, Cape Thompson, and Area 10

Livermore scientists and AEC technocrats typically understood Cape Thompson as an abstract space-to-be-engineered, the site for a spectacular demonstration of the efficacy of nuclear earthmoving technology. Pursuant to that objective, others, such as AEC ecologist John Wolfe, chair of the Committee on Environmental Studies for Project Chariot, saw it chiefly as the setting for a biological and ecological experiment. But these understandings of place, and the visions of a scientific and technological transformation of nature that underwrote them, were met with organized opposition in Alaska by those who saw the same place very differently: as hunting grounds, as home, and—for some environmental scientists—as a complex ecological system about which too little was known for a "controlled" experiment in nuclear excavation to be called, categorically, safe. But under whose authority were the risks posed by the Project Chariot nuclear harbor project to be defined? Who spoke for the ecosystem of Cape Thompson and places downwind, and for the safety of the people who lived there? How, and by whom, could competing claims to truth be judged, and on what bases were they to be evaluated? Scientific, political, and legal authority to carry out Chariot, and crucially, the power to prevent it, depended on which groups could make *their* claims to the imagined geographically engineered future of the Cape Thompson region authoritative.

As Chariot continued to take shape, a politics of opposition was built around two interrelated controversies. First was the question of whether or not the experiment could be conducted without harming the biological environment and especially the Inupiat Eskimos living in the region. A second issue concerned who held legitimate authority over the land (and land use) around Cape Thompson: the Eskimos from Point Hope Village or the AEC? For dissenting scientists and Eskimos both, effective opposition would depend on reconfiguring the power relations that had brought the experiment to Alaska in the first place; only by bringing new actors and groups into these nascent oppositional movements could the claims of Livermore's geographical engineers and Wolfe's Committee on Environmental Studies be effectively resisted. For both the politics of land tenure and the politics of scientific truth, local opposition to the AEC had to expand in scale if Chariot was to be prevented.

The resistance tactics of dissenting scientists in the AEC's bioenvironmental program and the Inupiat people of Point Hope both worked, in different ways, by challenging AEC claims to authority over the Cape Thompson region and by shifting the boundaries for dissent that the AEC had set in place around Chariot. The human geography and human ecology of the Cape Thompson region were central to the opposition of both the scientists and the Eskimos, so this chapter begins by examining the work of geographer Don Foote as he took up residence in Point Hope for his part in the Chariot environmental studies. Next I examine how networks of opposition emerged to contest the experiment, resistance that was organized, on the one hand, around oppositional land use claims stemming from Point Hope, and on the other around what I will call the oppositional science of Foote and several of the University of Alaska contract environmental scientists. Finally, we will see how the crisis of widespread national criticism of the project was resolved, by Plowshare's program builders, through another geographical move: the retreat to the pre-established experimental space of the Nevada Test Site for the first of Plowshare's nuclear cratering experiments, Project Sedan.

Geography in the Field at Point Hope

One of the scientists most incensed about the committee's statement of January 7, 1960, was geographer Don Charles Foote. Under his AEC contract, Foote had initiated an extensive study of Eskimo hunting patterns in the region. Because his studies were to provide essential data on Eskimo hunting as well as food consumption, his work was critical to the evaluation of possible risks associated with Chariot and thus, for the AEC, for ensuring the safety of the project. To Foote's dismay, however, the 1960 committee statement bore little relation to his preliminary findings. Had the committee even read his preliminary report, and if so, why had it been ignored? While the significance of Foote's fieldwork to the eventual outcome of Chariot might have seemed unlikely to senior Plowshare scientists as his research proposal passed through AEC channels in the spring of 1959, it would not be possible to ignore it forever.

With a passion for the northern latitudes acquired in the course of summer work on a U.S. Weather Bureau ship off of northern Greenland, service as a research assistant on an Office of Naval Research Russian-Norwegian boundary study, and two years of postgraduate research at the Geografisk Institute in Oslo, Norway, Foote had entered McGill University in Montreal in 1957 to pursue his interests in Arctic geography and northern studies. He earned a master's degree there in the Geography Department in 1958, but soon after embarking on a Ph.D. program in the same department, Foote became disenchanted by what he saw as an overly bookish, rather than field-based, approach to the region on the part of the faculty. After flunking his Ph.D. comprehensive examinations in 1959, perhaps willfully, he began a search for funding to support fieldwork in Alaska as an independent researcher and wrote the "Alaska Rural Development Board" to inquire about its summer research needs. The Alaska *Resource* Development Board, as it was actually called, advised him to contact Gerald Johnson, the test director at Livermore, regarding a "catalogue of resources" in the Cape Thompson area in connection with an unspecified Operation Plowshare.[1]

Foote jumped at the chance. He introduced himself to Johnson in a March 24 letter: "As a student of geography, I am primarily concerned with regional studies, planning and development in the northern fringe areas of the world." Foote went on to describe his field experience in the Arctic and his knowledge of "microclimatology and regional climatology, geomorphology and the processes shaping the landscape, soils and to a lesser extent vegetation."[2] Six weeks later, as the committee was organizing the nature and extent of its pre-Chariot studies, John Wolfe telegrammed Foote, asking for a complete proposal and budget. As Wolfe explained it, the biological program was concerned primarily with assessing the biological costs of nuclear excavation by relating preshot baseline data to postshot data. "While we anticipate no irreparable damage to ecosystems, no such study has ever been accomplished prior to an atomic explosion," Wolfe wrote. "Moreover this area is essentially in equilibrium, providing a much more substantial opportunity for basic studies than more 'civilized' areas."[3]

In his subsequent proposal for "Human-Geographical Studies in the Cape Thompson Area, Alaska," Foote improved on his earlier, rather vague notion of "regional studies" to mesh better with the outline of contemplated studies that Wolfe had sent him. His proposal for Phase I of his research, from May 1959 through May 1960, included as its objectives:

1. To locate on a seasonal or other appropriate time scale that human population which could be influenced by Project Chariot. Loca-

tion, movement, destination-origin and motives for movement will be indicated for all seasons.

2. To determine, by historical research, seasonal patterns of population location and activity not otherwise attained by a one year field study. . . .
3. To study the relationship between human economy and the biological environment. Seasonal diet-preferred and alternative-housing, transportation means, clothing etc., will be recorded both in field work and historical research.
4. To inquire into the cultural impact of project Chariot operations on those native people immediately adjoining the project site.[4]

Foote's proposal was quickly accepted. Taking leave from McGill, he became the senior investigator for human geographical studies, receiving approval for his requested budget of roughly $20,000 for the first year of the project. After attending a June organizational meeting for the program's contract scientists in Seattle, and after a summer of historical research at the northern studies libraries at McGill and Dartmouth, Foote took up residence in Point Hope on September 4, 1959 (figure 3.1).[5]

Figure 3.1. Don Foote driving a skin boat with an outboard motor near Point Hope, Alaska, in 1960 or 1961. *Don C. Foote Collection, Accession number 1996-113-3, Archives and Manuscripts, Alaska and Polar Regions Department, University of Alaska Fairbanks.*

Although Foote was, by all accounts, exceptionally bright, whether Wolfe and the AEC had selected this relatively inexperienced (and institutionally unaffiliated) researcher for studies of Point Hope and other nearby villages because few applications for the hastily organized human ecological studies had been received, or perhaps because they believed his findings could be easily ignored or manipulated, cannot be easily discerned. What is clear is that Foote's fieldwork in the Cape Thompson region (see figure 2.1), where he would reside nearly continuously with his wife, Berit Arnstad Foote, for three years, stands as a remarkable example of seat-of-the-pants scientific improvising, reflecting a positively Humboldtian commitment to integrated data collection. Foote began to accompany Point Hope hunters on their extended expeditions over land and sea, eventually learning to build and operate his own dogsled, train sled dogs, and to hunt for caribou, elk, and seal. Though his proposal had stated that he would spend the winter preparing his reports from Fairbanks, by October 23 Foote had changed his plans: "Only after speaking with you in early September," he wrote to committee secretary Ernest Campbell, "and with my first informants here at Point Hope did I realize that not only was I free to reside here for the year but that it was absolutely essential."[6]

Once ensconced in the field, Foot widened the scope of his data collection considerably; only by gathering as much information as possible about the relations between human economy and the physical and biological environment, he came to believe, would it be possible to make a realistic assessment of the hazards posed to the Eskimos by the Chariot experiment. He recorded and mapped, for example, the length and nature of hunting expeditions originating in Point Hope, the kinds of animals killed (which included various species of seal, walrus, whales, caribou, elk, various birds and fish, and to a lesser extent, polar bears and fox), when and where they were killed, the weight of the animals, the manner in which they were consumed (mainly raw, raw frozen, or boiled), and also the parts of the animal consumed (including bone marrow, inner organs, fat, and so forth). Not only food-gathering activities, which included the collection of bird eggs from nests along the coastal cliffs and a variety of grasses, but food purchases and the cash economy of Point Hope in connection with seasonal wage labor were recorded as well. He measured temperatures, wind speeds, and assessed the condition and movement of sea ice, ocean tides, and currents, all of which were related to hunting and fishing patterns and other human activities. Foote also collected on tape extensive interviews with the people of Point Hope concerning the relationship of Eskimo culture to seasonal environmental conditions, their selective adaptation of new technologies, and the history of the village (which, located on the Tikiraq land spit, was among the oldest continuously inhabited settlements in North America).[7]

All of this work depended a great deal on cooperation from his research subjects, for according to Foote's methods of participant-observation, the Eskimos acted as much the makers of new knowledge as they were the objects of a putatively expert analysis. Their willingness to have their hunting hauls tallied, and to allow Foote to tag along on the hunts and to inquire about their ways of life, depended on some level of personal trust in Foote, who had explained to the people what he knew of Project Chariot on his second night in the village. "It is my personal opinion, and that of the AEC," Foote had told them at that time, "that if Chariot proves harmful to the local people it should not be carried out."[8] But exactly what would it take to prove Chariot harmful before the experiment was even conducted? Foote, of course, had no access to AEC classified information regarding the nature and extent of expected radioactivity hazards from the proposed blasts; it was the AEC and its radiation specialists who would evaluate Foote's data, rather than the other way around. Foote's detailed spatial and seasonal observations of Eskimo hunting patterns and diet in relation to local environmental conditions soon provided the committee with a new set of "biological information" to consider.

What Foote observed was a complex, seasonally shifting system of relations between hunting and gathering activities, climate, and biosphere. Most significant, Foote quickly established the importance the entire Cape Thompson region to the Eskimos' minimal subsistence needs, given the large amounts of meat and fat that they consumed and used for feeding their dog teams and heating their homes. On the basis of his observations of hunting and consumption, Foote argued that "if the daily need for meat is figured at two pounds per pulling dog and per person over 14 years of age, Point Hope requires at least 7000 pounds of meat each week for minimum living conditions." This tremendous harvest of meat required of the Point Hope hunters, which depended largely on their own forms of environmental knowledge, developed by the Inupiat over the course of generations, not seasons, and based on observations of animal migrations in relation to local environmental conditions, was only possible by traversing large expanses of land and sea, and this depended on extensive dog team and boat travel. In slightly more than two months in the autumn of 1959, for example, Foote showed in his map of dog-team routes that "Point Hopers traveled over 10,000 miles in the area" (see figure 3.2). From September 1959 to October 1960, Foote reported that at least 402,265 pounds of meat were brought into Point Hope. Of this, 150,500 pounds were gathered within twenty miles of the proposed ground zero, while fully 400,000 pounds came from within forty-five miles of ground zero. Over 100,000 pounds of caribou meat was taken from the proposed (landward pattern) fallout sector.[9]

Foote stressed in his reports that these substantial quantities of meat were part of a rather fragile subsistence economy. Even with the benefits of technologies

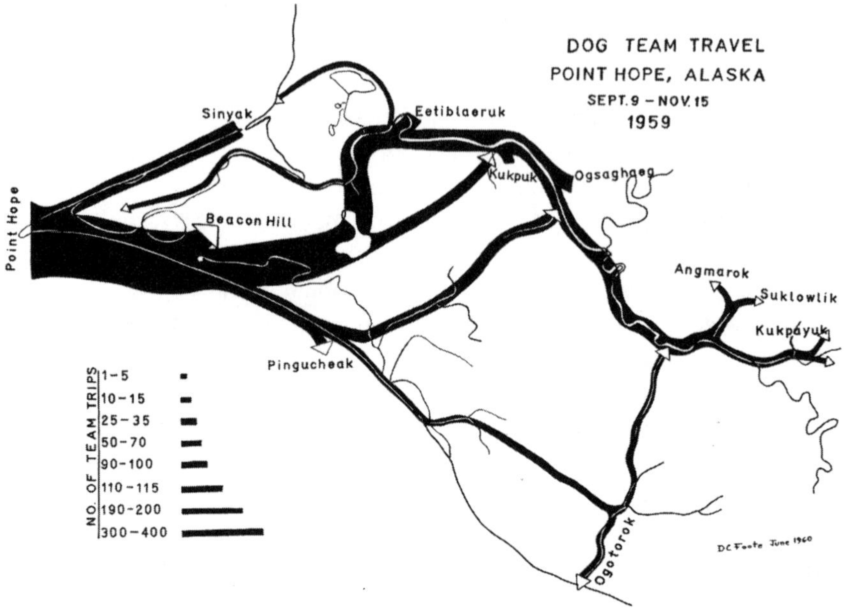

Figure 3.2. Don Foote's map of fall 1959 dog-team travel and intensity of trips by Point Hope hunters. *Source: Don Foote, "The Eskimo Hunter at Point Hope, Alaska, Part I: September, 1959, to May, 1960." DOE doc. NV0173276, Nuclear Testing Archive, Las Vegas, Nevada.*

such as rifles, binoculars, and outboard motors for boats, life in Arctic conditions at Point Hope was reproduced only on a "season to season" basis with little room for shortfall. Foote concluded his June 1960 report:

> Point Hope is now, as it always has been, a village oriented to subsistence living. In autumn, Kukpuk River fishing and caribou hunting prevent an undue drain on meat stored since spring, and aid greatly the returned summer laborer who may have foregone the important June seal hunting. During mid-winter, maximum hunting does not fill the daily meat demand. From late February to late June, the village either achieves a surplus meat supply, which brings minimum living conditions for the coming year, or it fails in the late winter-spring hunt and suffers hardship. Earned income and other cash receipts may prevent disaster, but the greatest privations are caused by wants which money cannot satisfy. At no time can the village long forsake its hunting activity, and yet remain confident of the future.[10]

As Foote illustrated in a map of winter and spring caribou and whale hunting, these important hunting seasons depended heavily on the land south and east of Cape Thompson, in part because heavy winds kept the area relatively snow free, making it an attractive region for caribou grazing. On the basis of these data, Foote made the case—in what seems an unambiguous response to the committee's January 7, 1960, statement favoring a spring detonation—that

> any event which would prevent the Point Hope hunters from reaping a harvest during any month of the year, but most especially in those periods from mid-January to early July and from September to December, could bring profound and prolonged hardship to the village. Although inland and coastal hunting can be extensive during late July and August, it is of the least importance to the over-all welfare of Point Hope Eskimo inhabitants.
>
> Of all months in the year, March and April are in many ways the most critical for the residents of Point Hope. *This is the period when inland hunting activity in and near Augotorux Creek, and coastwise traffic over Augotorux Creek, can be and usually are at their annual maximum.* In addition, it is a season when extensive sea ice hunting takes place for distances of ten to hundreds of miles westward from Augotorux Creek.[11]

Although this report had not been available to the committee in time for its December 10, 1959, meeting (at which the January 7, 1960, statement was formulated), Foote had submitted a preliminary report to the committee in advance of that meeting.[12] At that time, at the close of his first season on Cape Thompson, Foote had argued that "the core of the village is composed of individuals and families to whom this place and the manner in which they sustain themselves reflects who they are and what they are. An integral part of this identity is the food they eat, the manner in which it is prepared and the manner in which it is obtained. . . . These factors should be borne in mind as future plans are made for the continuation of Project Chariot. . . . [D]isruption must not proceed to the extent that substitutes for this way of life must be provided to sustain their well being."[13] Eskimo culture and identity, as Foote understood them, were rooted in place and the biological environment at the most basic levels. And while Foote's reports to the committee were characterized by a strong commitment to observation, measurement, and careful analysis based on available evidence, inseparable from his scientific work was what might be called an explicitly moral position: Foote believed that any loss of autonomy by local Inupiat over the conditions of their existence would necessarily do them harm. For Plowshare scientists and engineers, however, who had earlier that year enthused at Livermore's Second Plowshare Symposium of their ability to blow fallout over "uninhabited" lands east of ground zero (see figure 2.4), neither the vast, dynamic geography of hunting, gathering, and consumption that Foote described nor Foote's insistence on the interconnections among Eskimo hunting, food, and cultural identity was the kind of biological information that they wanted to figure into their evaluations of Chariot's feasibility. Although he did not put it in so many words, there was a strong sense to Foote's reports, perhaps best epitomized by a detailed map of Eskimo place names included with his December 1960 report, that he had come to believe that it was the *Eskimos'* geography with which Livermore planned to experiment.

"Point Hope Says No"

When Don Foote had arrived at Point Hope, he found that few people in the village had a clear sense of what Project Chariot was. Some thought the recently constructed base camp on Ogotoruk Creek was to be a new military site, perhaps a radar station. Others, however, were already aware of the nature of the nuclear harbor project, and also had, as Foote described it for one confidant, "quite a profound knowledge of fallout, the 'bomb' and 'what happened to those fishermans,'" the latter comment apparently in reference to the 1954 Bravo test in the Pacific.[14] Foote spent his first night at Point Hope explaining what he knew of Chariot and the bioenvironmental program to Dan Lisbourne, president of the Point Hope Village Council, and when he spoke to the village on his second night, he addressed 85 to 90 percent of the resident population, with Lisbourne translating into the Inupiaq. No one from Livermore or the AEC had yet bothered to visit Point Hope, but with Foote's arrival, the people of Point Hope were at last informed collectively of the essentials of the planned experiment—at least those essentials that had been made public—and they also learned something of how the Chariot Environmental Program, in general, and Foote's work, in particular, were to fit in with the project. However the people of Point Hope may have felt about being placed under Foote's scientific observation, the "100% cooperation" that Foote reported receiving for his studies was hardly surprising; if they trusted Foote, then it was in the best interests of the village for every last caribou, whale, or seal hauled in to be weighed and recorded by the visiting geographer.[15]

But Lisbourne, for one, did not pin all his hopes on science (in the form of the AEC's bioenvironmental studies) as the only means of preventing the nuclear-blasted "model harbor" on the Eskimos' traditional hunting grounds. He wrote to new Alaska senator E. L. "Bob" Bartlett to express his opposition to the nuclear harbor, and Bartlett, in turn, advanced the issue to the AEC.[16] At the very least, it was becoming evident to the AEC that its public information plans for Chariot must now take the local population into account. So the AEC's Campbell responded by suggesting that Foote meet again with the village council at Point Hope. "Your effort to allay the anxiety of the natives,"

Campbell wrote, "would be greatly appreciated."[17] Foote did request an audience with the village council, but he also made it clear to Campbell that he was nobody's lackey; he offered to speak about Chariot only "within the limits of [his] own knowledge," and added that the failure of any representatives from the AEC to visit Point Hope had been noted by some in the village who "were angered by speechs [*sic*] and lectures in Fairbanks, Juneau, Nome and Kotzebue, hundreds of miles from Ogotoruk Creek, while not one AEC representative visited Pt. Hope, just 31.5 miles from ground zero."[18]

Foote spoke to the village council on November 28, 1959. According to the minutes of the meeting, which were signed by the council and forwarded by Foote to the AEC, Foote again described how his work and the bioenvironmental program were related to planning for the experiment, and he read a prepared statement from the AEC avowing that no decision whether to proceed with Chariot had yet been reached, pending the outcome of further studies. Foote was asked questions about the dangers of atomic explosions by council members—questions which he felt he was in no position to answer—and he was also asked just why the project was so important to the United States and its AEC. New village council president David Frankson asked Foote to stress to the AEC "that he eats two good meals of meat every day and [to] show that the animals are mostly depend[ed] upon by the people here at Point Hope."[19] The council also requested that the AEC send a representative to Point Hope to explain the project in greater detail. After Foote left the meeting, the council decided to make a petition to include with Foote's letter to the AEC. The petition was also sent by the council to the Juneau office of the Bureau of Indian Affairs, and was eventually circulated within the Alaska state government. It stated:

> We the . . . Point Hope Village Council do not want to see the explosions at near areas of our village Point Hope for any reason and at anytime.
>
> We urgently objected to the plan of Atomic Blast at Cape Thompson.
>
> We are living only thirty miles west of Cape Thompson at Point Hope village, and as we still wants to continue living in Point Hope many years to come, the majority of Point Hope village community objected to the plan of Atomic explosion at Cape Thompson, we are asking for a fair settlement of this plan before it is done.[20]

But it was added to the council minutes that "if the U.S. Government and AEC can assure the village that it will not be harmed in any way, by loss of hunting grounds or changes in the way of living, the council will reconsider."[21]

The village council thus sought to turn the AEC's geography of authority on its head by shifting the burden of proof to the AEC, while insisting that the authority to consider arguments for Chariot's feasibility rested not in Livermore, San Francisco, or Washington, D.C., but in Point Hope.

The AEC responded to the protest almost six weeks later, indicating that they would send a representative to speak at Point Hope before the end of March 1960. In that month, following a series of meetings in Juneau and Anchorage with state officials and environmental scientists, an AEC team was sent for a public information tour of the northwest Alaskan villages of Nome, Kotzebue, Noatak, Kivalina, Point Hope, Point Lay, Barrow, and Shishmareff—almost two years after the first drilling for core samples on Ogotoruk Creek—to explain Project Chariot to the Eskimos.[22] But instead of sending a radiation biologist or medical physicist to discuss the evaluation of radiation hazards from Chariot, the AEC sent two representatives from its San Francisco Operations Office, public relations specialist Rod Southwick and head of technical operations Russell Ball, along with Robert Rausch, an environmental studies committee member and parasitologist who was familiar to many Eskimos in the region from his work with the Public Health Service. The team barnstormed from village to village by propeller plane, apparently without advance warning to their hosts, with a reel of film in hand: *Industrial Applications of Nuclear Explosives*, the motion picture simulation of the nuclear harbor excavation that had been prepared two years earlier for the peaceful uses of atomic energy conference in Geneva.

At every village before the team reached Point Hope, neither objections to Chariot nor any "significant questions" were posed after the oral presentation and movie, according to the team's report.[23] And yet, as historian Dan O'Neill points out, the meetings must have been beset by communication problems, for not only did a minority of Eskimos in the region speak English at this time (and even then it would have been as a second language), the technical jargon employed in the AEC team's explanations would likely have made clear communication problematic for those who did understand English.[24] As George King, a Bureau of Indian Affairs teacher, recalled the meeting in Noatak in response to a query from Foote: "Yes, we had the 'propaganda' team here too, but they hit so fast and so early nobody really woke up until they were gone. They arrive about 8:30 A.M. and asked if they could meet with the village. . . . [After showing the film] The 'mouthpiece' then asked if anyone had any questions, and after a brief pause he assured all that 'the U.S. Government would never do anything that would hurt them in any way.' With that they swiftly departed . . . as a group of sleepy eyed, half awake Eskimos were wondering 'what they came for.'"[25] Later that afternoon, however, as "the mouthpiece" and his colleagues arrived to provide similar assurances in Point Hope, where many had been waiting several months for this meeting, they received a much different response.

As he had in the other villages, Southwick assured the roughly one hundred people in attendance at the Point Hope meeting that no one would be harmed by the Chariot experiment, and that "their hunting and fishing would not be interfered with. This is very important to the Commission."[26] The audience at Point Hope grilled the AEC representatives for fully three hours with questions pertaining both to radiation safety issues and to the rationale behind the choice of Ogotoruk Creek as the location for the experiment. The Eskimos stressed the importance of the project site to their caribou hunting and springtime crowbill-egg gathering, and they raised questions about radioactivity-related diseases in Japan and the Marshall Islands. The Reverend Keith Lawton, an Episcopalian priest, raised additional questions ranging from the AEC's considerations for dealing with permafrost to ground shock hazards to the expected release of radioactivity from the nuclear earthmoving explosions.[27]

In response, Southwick and Ball pressed Rausch into service answering questions about fallout, a matter about which Rausch was the first to admit he had no special knowledge. His answers were, by Don Foote's recollection, reiterations of what environmental studies committee radiation specialist Allen Seymour had recently told project scientists in Anchorage, and in this sense, Foote later reflected, Rausch was a "duped but sincere man."[28] Duped or otherwise, as the "propaganda team" was put on the defensive, Point Hope was given a good deal of misleading information about the biological effects of Chariot. For instance, Rausch maintained that "some radioactive elements would be gone in a matter of hours, while others would take longer"; this was true, but highlighting only the hourly time scale obscured the fact that the radioactive half-lives of potentially dangerous radioisotopes like cesium-137 and strontium-90 would take more than thirty years longer to decay.[29] In addition, the team claimed that studies of cattle in Nevada and fish in the Pacific provided evidence that fallout would not cause harm to fauna in the Arctic, a rather speculative ecological comparison with no basis in actual evidence. Still other answers provided by the AEC representatives would be better characterized as outright lies than misinformation; as a professional working for the AEC, Ball surely knew better when he insisted that "at this distance [the thirty-one miles from ground zero to Point Hope], the amount of airborne radioactivity which could reach here could not *possibly* be enough to cause injury to the people or the animals. There's just no chance of that."[30]

Despite Southwick's assurances that Chariot would not be conducted if it would harm the Eskimos or interfere with their way of life, and regardless of the team's appeals to the authority of science as the ultimate evaluator of the risks posed by Chariot—that is, that studies in Nevada and the Pacific had shown that fallout had not caused harm to wildlife, and that the bioenvironmental studies would prove the experiment safe for Cape Thompson—the Point Hope Eskimos were not easily convinced, nor did they see any particular merits to the model harbor project. One resident of the village, Kitty Keenvevaek, was undoubtedly speaking for many when she put it to Southwick, Ball, and Rausch, "I'm pretty sure you don't like to see your home blasted by some other people who didn't live in your place like we live in Point Hope."[31] The meeting ended when, after about two minutes of silence, an Eskimo woman in the back of the room shouted, "Point Hope says 'No.'"[32] When the AEC team departed, the village council once again declared its unanimous opposition to Chariot.

Toward an Oppositional Science

In early March 1960, two months after the confrontational meeting in Point Hope, the AEC took up the question of whether or not to proceed with the next phase of Chariot "preshot" research. Moving forward with Livermore's nuclear earthmoving experiment, and with the environmental program riding on its coattails, required the assent of the AEC in Washington. Basing its decisions chiefly on the preliminary (January 7, 1960) report of Wolfe's Committee on Environmental Studies and the recommendations of its newly formed Plowshare Advisory Committee, and evidently with little regard for the concerns of Point Hope villagers, the AEC announced in a March 4 press release that "nuclear detonations could be conducted under certain limitations," but that "continuing study during 1960 is required for firm recommendations." Environmental surveys would thus continue throughout 1960 "to assist the [Atomic Energy] Commission in determining whether conduct of this experiment would be feasible on the basis of engineering and public health and safety data." Yet it was also noted in the statement that the explosive design currently planned by Livermore would now include only a single 200-kiloton and four 20-kiloton explosives detonated simultaneously. This further reduction in explosive yield (from 480 to 280 kilotons) was not explained, except to say that the longer "row-charge" would produce useful data for future excavation projects such as canal construction. Laboratory experts, it was added, anticipated that more than 90 percent of radiation produced by the explosions would be confined underground, though "other calculations made on the containment of radiation from nuclear explosives buried in the earth range up to 99 percent confinement."[33] Exactly how much radioactivity was expected to be released in the blasts, which is to say, 90 percent of *what,* was not disclosed, nor were the bases on which these predictions—for an experiment unprecedented in both its scale and its medium of permafrost soil—were premised.

Shortly before Southwick and Ball's tour of the Inupiat villages of northwestern Alaska, they had met with investigators in the bioenvironmental program in Anchorage, along with committee chair Wolfe and the radiation biologist Seymour, to discuss the previous field season and expectations for the 1960 season. Although the cultural differences were not so great as those that would be experienced by the AEC team days later at Point Hope, this meeting too was characterized by open challenges to AEC and committee claims, but here the terrain of conflict was not only Cape Thompson's tundra ecosystem but also the experimental space that had been erected around Project Chariot. The Anchorage meeting was the first chance for contract scientists in the environmental program—from the Alaskan Division of Fish and Game to the University of Alaska biologists to the independent contractor Don Foote—to question committee members in person about what some had seen as the distortion of their data (if not utter misinformation) that had been advanced by the committee in its preliminary statement. By all accounts, the critics of the Chariot enterprise did not pull their punches.[34] Several of the biologists harangued the AEC and committee representatives about what seemed a decidedly nonobjective interpretation of their data. Foote expressed concerns, as he described the meeting later for one correspondent, about the "honesty of the [Atomic Energy] Commission" in its relations with the Eskimos, and openly questioned the "sincerity" of the environmental program.[35]

Such remonstrations over scientific honesty, sincerity, and objectivity were, in the field of scientific ethics, no small charges. The AEC team responded by insisting that the January 7 statement "was based on incomplete information and further investigation is required prior to making a final recommendation to the Commission."[36] The dissenting bioenvironmental scientists—including botanists Les Viereck and Albert Johnson, mammalogist Bill Pruitt, and ornithologist Gerard Swartz of the University of Alaska, biologist Jim Brooks of the Alaska Department of Fish and Game, and the geographer Foote—left the meeting under the impression that errors in the January 7 statement would soon be corrected. Even after a March 13 *New York Times* article parroted some of the most disputed points in the statement, including claims that spring was the preferable season for the Chariot blast because hunting activity in the area would be minimal at the time while the snow cover would be high, the contract scientists still chose not to publicize their dissent.[37] But their patience was running out.

On June 8, 1960, with no corrections yet issued, Johnson, Viereck, Pruitt, and Swartz, the four University of Alaska biologists, submitted a letter of protest to the committee concerning the January 7 statement and its dissemination. In language that could not have been lost on Wolfe or any other committee scientists, the letter attacked what seemed to the biologists to be the subjective, unscientific nature of the committee's conduct, stressing, for example, that the statement's claims about winter and spring conditions around Ogotoruk Creek were based "largely on opinion rather than observation." The letter continued: "We had hoped and expected that the statement would be corrected following the March meetings. This has not been done. We wish, therefore, to formally register our disapproval of it, (1) because of the lack of scientific objectivity in the statement, (2) because of the misinformation it contained, and (3) because it implies that we have furnished data to support it."[38]

The biologists were particularly upset about the statement concerning snow cover in the downwind region of the proposed landward fallout sector. Chariot's experimentalists had posited that long-living radionuclides would flow more or less harmlessly out to sea with the spring and summer snow melt, but the Fairbanks biologists noted that, according to their observations, snow cover in the spring was in fact "very sparse due to winds almost continually in excess of 20 m.p.h." Indeed, the sparse snow cover was what made the Ogotoruk Creek region so attractive to caribou grazing and, in turn, to Eskimo hunters during the spring season, but the presence of caribou in the region, the biologists complained, had also been obscured by the committee. In addition, the biologists suggested that a full-scale investigation of marine mammals would also be necessary, given the fact that both landward and seaward fallout sectors had been deemed "acceptable" in the statement. But how could either fallout sector be classified as such so early into the study? The four biologists noted: "The scientific investigations will provide many answers to some of these questions, but in the meantime, it is unfortunate that the Committee felt it necessary to offer opinions on scientific questions for which it had little or no special information. We sincerely hope that another statement correcting the information . . . will be released by the Committee following the June review of the interim reports." They added, "continued Committee support of the January 7 statement can only lead to embarrassment of the Committee and to the individual investigators as contradictory information are made public."[39] Beyond the substantive issues raised in the letter, here was the crux of the matter: the biologists *would* take their contradictory information to the public, and if the committee did not take their data into account, then it would soon have a public controversy on its hands.

Soon after, Don Foote submitted a separate statement to the committee formally requesting that it withdraw its claim that, during the spring, "local hunting activity on the land is low."[40] But he expressed stronger criticism, along with a clear warning, in a letter to Wolfe himself. "Naturally," he wrote, "I will not allow the results of my research work to be misused." For Foote, the AEC's use of misleading claims about Eskimo hunting patterns as a justification for the spring blast suggested one of two things: either the committee was not a "sincere body," or it did not consider his research valid. But like the University of Alaska biologists, Foote understood well that public criticism developed from among the AEC's own contract scientists could be far more damaging to Chariot than opposition from "outside" scientific actors. "Participants in the biological program must have confidence in the Environmental Committee," Foote cautioned Wolfe. "If they do not, the AEC will have a group of opponents more potent than imaginable."[41]

In response to these challenges, John Wolfe took up the task of both pacifying the dissenting investigators and clarifying, as he saw it, the extent of their authority over Chariot matters. "Thank you for your interesting letter," Wolfe began his reply to the Fairbanks biologists, and he informed them that a new committee statement was under preparation that would be forwarded to the scientists at the project site. But he continued by stressing the committee's role, rather than that of its contract scientists, as the evaluator of information from the environmental studies:

> We must evaluate data from many sources—more than 30. These must be arrayed against best available estimates of the distribution of ejecta including levels of radioactivity. . . . As for snow and ice. . . . Note that it was not said, nor will it be, that snow cover is essential to safe excavation; nor is ice, which covers most of the sea at this period of the year.
>
> As for marine mammal studies, the program is undergoing expansion and we are pleased with developments. I note that none of you is under contract in this area, but of course, your interest is appreciated.
>
> We are aware that not all scientific problems will be solved in this first effort of biologists to collaborate in a technological program of this magnitude. That we are in it at all is a unique and distinguished achievement. . . . From this position let us proceed constructively and patiently to the end that biology and the environmental sciences *are*, and need to be, part of the warp and waft of the endeavor involving peaceful uses of atomic energy.[42]

Or, as Wolfe explained to Foote in another bit of boundary policing, "All competencies must be joined in this venture including those of the Committee and its bioradiological advisors."[43] No correctives to the January 7 statement were ever issued.

BIOLOGICAL OBJECTIONS

On August 15, 1960, in an event planned explicitly "to counter any adverse publicity to the proposed Plowshare experiment," the AEC arranged for journalists to visit the project site at Cape Thompson to listen to Wolfe and Seymour discuss the findings of the Chariot environmental program.[44] Although many environmental scientists were at the time stationed in the field at Cape Thompson, only committee members Wolfe, Seymour, and Norman Wilimovsky were available to the press at this time; Viereck and Pruitt, most notably, were working at their study sites and were not aware of the media visit until after it was over.[45] This was no oversight. In his introduction of the environmental studies program to the national media, Wolfe was absolute in his disregard for the reservations expressed by Foote and the dissenting biologists. As the *New York Times* reported on the press conference, in an article headlined "Proposed Atomic Blast in Arctic is Called Safe," Wolfe announced that a fifteen-month field study costing two million dollars "had produced no evidence that the detonation would damage the Eskimos' relationship to their environment and livelihood." Wolfe went on to pronounce for reporters "that there are no biological objections to the shooting on the basis of our investigations. Obviously, with a neighbor as close as 180 miles away there may be problems of which we are not aware, and which you would have to take up with the State Department."[46] The reference to the Soviet Union, the closest international neighbor to Cape Thompson across the Bering Strait, amounted to more boundary work by Wolfe, whether by intention or not, neatly separating the realms of science ("no biological objections") and politics (the territory of the State Department), even as Wolfe's own politicking rendered such a division between scientific and political practice problematic. For whatever the contemporary international political context—including tensions over the ongoing test moratorium—Wolfe more than anyone knew that the problems confronting Chariot were not occurring only at the international scale. For his Committee on Environmental Studies, anyway, the political difficulties experienced were the apotheosis of local problems: the emergence of challenges to the committee's authority were rooted in the views of the people who lived near Cape Thompson, and in questions over how the biology, ecology, and human geography of the place itself were to be evaluated and by whom.

The *Times* article pronouncing Chariot safe also touched on, but essentially obscured, opposition to the project from Point Hope. Registering dissent from town "postmaster" David Frankson—without mentioning that Frankson was also the president of the village council—but also quoting an unnamed Point Hope resident who "shrugged off worries of fellow townsmen as silly," the *Times* reported that Point Hope was "divided" on Chariot.[47]

Wolfe's claim that there were no "biological objections" to Chariot carried the apparent weight of thirty separate environmental studies, two million dollars' worth of scientific work, and the evaluation of a putatively independent committee of experts. But exactly what "biological objections" could there have been at this point? Investigators in the bioenvironmental program had begun putting together an inventory of the flora and fauna of the Cape Thompson region, including some baseline radioactivity data. Sampling plots had been established to estimate the quantity and variety of plant and animal species, and through analysis of stomach contents and droppings, food webs were being constructed to model energy flows through the ecosystem. Other forms of observation were employed to study sea mammals, birds, and fish, and light aircraft had also been utilized, to a limited extent, for tracking caribou herds in the region. The environmental scientists collecting these data, however, had not been told by Livermore how much total radiation, let alone precise quantities of particular radioisotopes, was expected to be released into the atmosphere. All of this was classified, "restricted data." Moreover, the evaluation of biological costs, based on predictions established with Livermore's nuclear cratering curves, was a task assumed by Wolfe's committee, on which radiation specialists Seymour and Kermit Larsen were both employed by AEC-supported laboratories. Any "biological objections" from Don Foote or the Fairbanks biologists would have been deemed, a priori, at once outside their areas of competence and beyond the limits of accessible information.

But the dissenting scientists could still take issue with how their data had been evaluated and publicized by the committee. The trio of Foote, Pruitt, and Viereck formed the core of what would become a jointly scientific and political opposition to Chariot over the next two years. Their criticism of the committee's methods of evaluation and interpretation and, subsequently, the trio's rerouting of the conduits through which their work was made known to the public and to other branches of government became a means by which to voice questions about Chariot that the committee had, to this point, refused to address. Only by transcending the boundaries of authority established through Wolfe's committee could their dissent become legitimate.

OPPOSITIONAL SCIENCE: A MORAL SETTLEMENT

In December 1960 Les Viereck informed University of Alaska president William Wood that he was resigning from the environmental studies at Cape Thompson. When he took the position in 1959, as Viereck explained, "I felt that the study would be purely biological in nature and would in no way become involved in the politics of the AEC." Subsequent events, however, had caused Viereck to rethink this assumption, and to reflect on the nature of scientists' responsibility to their work: "It has often been stated that a scientist working under a government contract should not worry about the interpretation of the data he is hired to collect. I strongly disagree with this attitude and feel that it is the duty of every scientist to protect his data and to be sure that it is interpreted correctly. A scientist's allegiance is first to truth and personal integrity and only secondarily to an organized group such as a university, a company, or a government. If we allow government agencies to interpret our data we are falling down in our duties as scientists."[48]

If Viereck's naïveté about the separation of politics and science under AEC contracts had been eviscerated by the actions of the committee over the previous year, his idealism about science as a search for truth, depending on the integrity of individual scientists, had not. After composing his case against the committee in his letter to Wood—from the committee's original January 7 statement to its failure to respond to the biologists' first letter of protest, and from its neglect of Don Foote's data to its misleading pronouncement of "no biological objections"—Viereck avowed that he could no longer "have his name associated with the project. . . . Because the Project Chariot study is not a purely biological study but one intimately connected with AEC politics and directed by an Environmental Committee that is biased in its decisions and membership, I wish to resign from the project. I greatly regret doing so because I realize that the biological knowledge gained by this study is of great value and that the study as a whole is one of the best ever conducted in an Arctic region."[49] In response to the letter, President Wood, who had little patience for academic politics of any kind (except those from the top down), was more than happy not only to accept Viereck's resignation from the Chariot studies but also to dismiss him from his teaching responsibilities in the Biology Department, a position that had been created for Viereck when he was recruited by University of Alaska biologists (from his doctoral studies in Colorado) to take part in the contracted environmental studies.

The AEC, for its part, commissioned a report of staff comments in preparation for possible inquires about Viereck's resignation from the project.[50] The report staunchly defended the integrity of the committee and attacked what it insisted was Viereck's misunderstanding of the nature of the biological program: "Members of the Committee have stated many times that some biologic damage will occur no matter what activity man undertakes. They have stated that even walking over the area to collect specimens causes some 'biologic damage.' There obviously will be total biologic destruction in the crater area. As Mr. Viereck points out in his letter, one of the objectives of the environmental studies is to assess the biological cost should Project Chariot be authorized."[51] Apart from its rather inane comparison of the "biologic damage" of footprints on a sample plot to concerns over the contamination of Eskimo food sources, the issue, for the AEC, was that one of its contract scientists had again misinterpreted the nature of Chariot as a bioenvironmental *experiment*—the radioactive tracer experiment writ large—the value of which could only be realized by comparing preshot to postshot data. Viereck had believed that the bioenvironmental program would have a role to play in evaluating the risks of Chariot before the nuclear blast was authorized, an understanding reached, of course, in response to many of the AEC's public statements indicating that the environmental program would help to assess the safety of the experiment before it could take place. For the AEC, the preshot studies meant one thing in public, and another in private.

Like Viereck, Don Foote had begun his field studies before having completed his doctorate. And also like Viereck, Foote had entered the environmental program holding to an ideal that scientific truth was something that resided outside the realm of politics. Indeed, he had assumed that the environmental program stood squarely on the side of science. As Foote recalled for Wolfe:

> One reason that I took the research position in Human Geographical Studies was my conviction in the sincerity of your personality and your dedication to the principle that the wise use of nuclear explosives demands considerations of pre- and post-detonation effects on the biosphere.
>
> My belief in you and the program has been shattered. To a great extent this was the effect of my politically immature mind awakened to the modern interplay of politics and science. It was caused by my realization that when science becomes directly involved with politics, the concept of scientific truth becomes modified, warped, and totally abandoned.[52]

Foote's awakening, however, did not lead him to withdraw to a purely scientific "side" of the modern interplay; on the contrary, his response was to make his science more political. But for Foote, along with Viereck and Bill Pruitt, any settlement of science with political activism could only work to the extent that it was based on doing "good science," on the one hand, and on an underlying moral and ethical context, on the other. For the three young scientists, the forging of an "oppositional science" to counter the AEC's misleading claims and duplicitous behavior was thus, in a sense, a spirited political defense of traditional notions of apolitical science.

Conceiving of their work *as* oppositional was a decisive step in this defense. From summer 1960 to spring 1961, Viereck and Pruitt each carried on an extensive correspondence with Foote at Point Hope, covering everything from the exchange of scientific data to commiseration over committee censorship to, ultimately, how to publicize their own findings and concerns about Project Chariot outside of the two agencies that had, to this point, muted their concerns: the committee—in effect, a panel of experts convened by the AEC's

Division of Biology and Medicine—and, through increasingly vigorous acts of self-censorship, the University of Alaska.[53] If, for Viereck, resignation from the environmental program was the best means of addressing his misgivings about the committee, Foote and Pruitt chose what might be called more subversive tactics: they believed that opposition to the committee's findings would be more powerful precisely because it originated from AEC-funded scientists like themselves. As Foote put it to Pruitt, "Simply said, we have been far too honest for our opponents." By continuing to report their findings through the committee, by being "wishey-washy" over Chariot, and by clinging to notions that objective science could be achieved under the conditions set forth by the committee, Foote believed that the dissident scientists, until now, had been playing into the hands of Livermore and the AEC.[53] The AEC, as Foote argued to Viereck, was a "highly organized group with far more experience in these matters than you." Through its procedures of classification and compartmentalization of knowledge, he claimed the AEC relied on a philosophy that "ignorance produces a favorable atmosphere" of:

(a) confusion among the opponents
(b) lack of united front from opponents
(c) greater freedom for policy changes from the top
(d) places the scientist in "the know" in the position of father protector and advisor.[54]

In response to these perceived tactics, Foote, Viereck, and Pruitt began to develop their own strategies for publicizing their findings and expressing their concerns over Chariot in an effort to undermine committee statements and, ultimately, to derail Livermore's plans for geographical engineering in Alaska.

To be effective in their opposition, the scientists would have to shift the flow of knowledge about the project and about Cape Thompson's regional ecology from the private communications of their AEC reports to the public space of the popular and scientific presses, and to establish new networks of communication with outside experts, activist groups, and state and national government agencies. But this was only possible, Foote, Viereck, and Pruitt believed, by doing better *science* than that of the committee. This meant presenting "raw facts and information" that "speak for themselves and raise questions so serious that no thinking person should disregard them."[55] Their work would demand "strict objective scientific attitudes" in order to highlight the uncertainties associated with the project and to expose committee malpractices and ethical misdeeds to the public and concerned agencies.[56]

To build an oppositional science that could succeed in challenging AEC authority, though, Foote believed that the dissident scientists would also necessarily need to expand their investigations beyond their fields of training. He argued to Viereck in October 1960:

> I am afraid I must admit our biggest stumbling block has been our unwillingness to do the fantastic job of reading in the field of radiation biology. *This must be done!* . . . Had anyone known the facts there would have been a storm in Anchorage [at the March 9, 1960, meeting]. Now it is here that LRL [Lawrence Radiation Laboratory at Livermore] will collapse. They cannot predict:
>
> a) the total amount of radiation to be released.
> b) the amount of fractionalization of fission products.
> c) the distribution of particle size.
> d) the height or nature of the atomic cloud and debris column.
> e) sudden changes in wind velocity and direction
> f) the redistribution of radioactivity due to blowing snow.
>
> Because of these reasons they *cannot* predict what the fallout will be. We must prove this. In addition, we must present a series of models, based on available knowledge, of what the varying doses could be in the area and how these can be related to knowledge already gained in the field of radiobiology.[57]

Foote thus enlarged his investigations well beyond his original proposal for investigating the human geography of the Inupiat Eskimos of the northwestern Alaskan villages. Yet he also knew that a successful oppositional science would require the evaluation of experts in the fields into which his work was expanding, and so he and Pruitt initiated contacts with numerous biologists, physicists, economists and anthropologists who were not affiliated with the AEC. The dissident scientists also struck up a correspondence with the activist Committee for Nuclear Information (CNI) in St. Louis. They contacted government authorities in the state of Alaska and the U.S. Department of Interior to keep them apprised of their research, alert them to their concerns, and to press for an independent review of Chariot.

DON FOOTE'S PROPOSALS

Another aspect of Don Foote's oppositional work was an effort to reconstruct and document the history of Project Chariot, at least so far as was possible using publicly available information and the findings of his comrades. He was especially interested in relating the transformation of AEC public information postures to the changing technical aims of the experiment, and here—as he sought to educate himself about the technical aspects of nuclear excavation—his investigations pointed back to Livermore.[58] Several of Livermore's technical reports on Chariot and nuclear cratering had been published, in accord with earlier plans for an "open" public information policy for Plowshare, yet when Foote requested copies of these reports from the AEC, the San Francisco Operations Office replied, "We have reviewed the scope of work for your contract and find that at present there are no publications we are aware of from this office or any AEC contractors which are required for fulfillment of your contract for Human Geographical Studies beyond those which have already been sent to you."[59] Given that Foote was living in Point Hope at the time, such boundary policing was tantamount to a kind of special classification, a need *not* to know. Foote's request for the minutes from committee meetings was rejected as well: "Because of their nature and purpose, these minutes are for Committee and Commission use only."[60]

But even as his own queries—and his own data—were at times rendered inadmissible by the committee, Foote's organizational history of Chariot suggested to him that, in its assumed role as the guarantor of public health and safety, the committee itself had in fact overstepped the authority granted in its charter.[61] That is, although the committee did have the authority to evaluate the environmental studies, and to make certain recommendations to Livermore and the AEC, it did not have the authority, Foote insisted, to judge whether or not the project could proceed safely. So Foote produced two proposals, to be read at the committee's December 6, 1960, meeting, which he submitted directly through committee (and USGS) geologist Art Lachenbruch, the only committee member whom Foote seems to have trusted by this point. The first called for an immediate public statement declaring "a redefinition of the committee function" to direct the environmental program, assimilate and review the findings of its investigations, recommend further research, and synthesize knowledge gained from the program. This statement must also clarify, Foote instructed, "that the function of the Committee is *not*, . . . [can]*not* and will *not* be" to assess the biological consequences of a nuclear explosion at Cape Thompson, nor to recommend the conditions under which such an explosion could be conducted. In addition, the proposal called for the minutes of all committee meetings to be availed to the public.[62]

If the first proposal did not sufficiently offend committee scientists, then the second one surely would. It called for the committee to consider disbanding itself, by majority vote, and to be restructured so as to assist in organizing the Chariot environmental program but not to evaluate the studies themselves and, in turn, for the committee to recommend to the AEC that "an independent fact-finding body be established of members of known professional distinction," to be appointed by either the National Academy of Sciences or the United Nations.[63]

Did Foote believe that either of his proposals stood a chance of passing? Surely not. Even the sympathetic Pruitt, Viereck, and Swartz thought the forceful tone of the proposals "might do more harm than good."[64] For Foote, though, it was a question of "moral obligations to the men" to give committee members one last chance before he went public with his claims. It had been a trying year for Foote. His studies had been systematically devalued or ignored in the committee's preliminary statements. Along with his requests for information, his requests for logistic support, including items as basic (and common at the Ogotoruk Creek base camp) as a portable generator for powering a Coleman stove to heat his home, had been repeatedly denied. Foote also, he believed, and as seems likely, became the target of FBI or CIA surveillance—as a rabble-rouser, presumably, who was not only opposing a major AEC project but also, perhaps, mobilizing Eskimo opposition to Chariot. The proposals were Foote's line in the sand. "When I draw and quarter them in print," he explained to Pruitt, "they will have nothing to say."[65]

Controversy

Although the internal protestations of Foote and the dissenting biologists had registered with Chariot advocates, they had done little to dampen the experimentalists' enthusiasm when the AEC revisited the matter in a February 1961 meeting in which it was to decide, once again, whether to proceed with Livermore's proposed demonstration project. A 7.9-million-dollar increase in the Plowshare budget was proposed, including nearly 5 million dollars allotted for "site preparation" in Alaska for Chariot, "to meet a shot schedule in the Spring of 1962." With strains on the nuclear test moratorium already becoming evident, AEC commissioner Carroll Wilson added that "the Commission should be prepared to proceed with Chariot as soon as the test moratorium is lifted by the White House."[66]

By this time, a Division of Peaceful Nuclear Explosives had been set up within the AEC. Although scientific, technical, and creative direction for Plowshare still emanated from Livermore, the move was designed, it seems, to allow the AEC to shift the program's financing, management, and public relations work away from the new division's more established counterpart, the Division of Military Applications. Such an administrative division would perhaps be better suited to supplying information and fielding queries about Plowshare projects as the program became more active in the run-up to what was to be, it was hoped, the first of many nuclear earthmoving ventures. John Kelly, an AEC technocrat who had built on two years of graduate training in mathematics and physics with thirteen years as a radar and instrumentation specialist in the

Navy Department's Bureau of Ships and the AEC, had been appointed director of the new division. It was Kelly who informed the AEC at the February meeting that he had spoken with several Alaska legislators and assured them that Chariot was a safe test.[67] Apparently no difficult questions were raised among AEC authorities over the safety of Chariot.

The following week the AEC's Plowshare Advisory Committee (PAC) "was unanimous in recommending the Commission proceed with Project Chariot at the earliest possible date." The PAC, composed of scientists and engineers drawn from industry as well as university science and engineering departments, based its evaluation chiefly on sources internal to the AEC: Livermore's predictions for the quantities and dispersal patterns of radioactivity produced by the explosives; the preliminary statements of Wolfe's committee; and the special meetings held with Livermore and Nevada Test Site scientists and engineers.[68] It is unclear whether PAC members were even aware of the concerns of environmental scientists in Alaska. What is clear, more generally, is that any dissent over Chariot—from scientists or from the Inupiat of Point Hope—was either obscured or ignored in official deliberations over Chariot in the first months of 1961. As spring arrived, however, new scientific reports on Chariot emerged in the effort, among critics of AEC practices in Alaska and elsewhere, to create a more public trial for the proposed nuclear harbor project than the AEC's internal proving grounds had allowed.

"'OBJECTIVE' BUT CONFLICTING REPORTS"

Shortly after Chariot received these endorsements from the PAC and AEC, Wolfe's committee completed its "First Summary Report," which, though claiming to consist of an evaluation of the progress of the bioenvironmental program "to date," inexplicably described only the work from the first field season at Cape Thompson, not the second. In any case, few of the findings from either field season of the University of Alaska bioenvironmental studies would have supported the report's claim that "the chance of significant biological costs . . . appears exceedingly remote."[69]

For the program's biological investigators, according to the Fairbanks Biology Department chair, Brina Kessel, the report seemed plainly biased in favor of conducting the nuclear excavation experiment.[70] Les Viereck, who was still in Alaska despite his resignation from Chariot and subsequent dismissal from the university, was surprised to find that his data on the relatively sparse snow cover of the Ogotoruk Creek region had been downplayed, seemingly in accord with Livermore's contention that fallout would be washed out with the snowmelt. Bill Pruitt found that all of his references to the potential concentration of strontium-90 in lichens, and to the caribou who fed on the lichens, were omitted, even though the references (which included citations to recent Norwegian studies of the concentration of Sr-90 in reindeer) primarily took the form of calls for further research.[71] And yet while Pruitt's comparison between northern ecosystems had apparently been deemed illegitimate in the First Summary Report, the mammalogist was infuriated by the committee's frequent references to biological and ecological studies conducted at the Nevada Test Site, including studies of a herd of cattle kept on the NTS grounds which were used to claim the safety of the uptake of radioactivity in large herbivorous mammals elsewhere.

The summary report's treatment of the human geographical and human ecological studies was equally selective in its representation of the investigators' progress reports. Don Foote, expressing his own evolving views of native cultural rights along with the transmission of his data, had written that disruption of Eskimo hunting patterns "must not proceed to the extent that substitutes for this way of life must be provided to sustain their well being." At no time, Foote had argued, could the Point Hope Eskimos forgo hunting for even one month without experiencing hardship. But the committee summary of Foote's findings contained a much softened conclusion, noting that "disruption of hunting activities and attendant harvests *without substitute* in any month could result in hardship to the Eskimo . . . in the opinion of one of the investigators in human geography."[72]

In March 1961—after the First Summary Report had been composed and circulated to the contract scientists but before it had been published—the *News Bulletin* of the recently formed Alaska Conservation Society (ACS) published a review of the Chariot biological program by Pruitt, of the botanical program by Viereck, who had become the president of the fledgling activist group, and a short history of Chariot and the Plowshare program by Foote. The review contained no classified information (the contract scientists had not been privy to any), but it did reprint most of Viereck's letter of resignation from the Chariot environmental program alongside editorials that more explicitly questioned the safety of the AEC's plans. One thousand copies of the thirty-page newsletter were printed (the distribution of the bulletin thus nearly quintupling the organization's membership), and it was mailed to scientists, government officials, conservation and wilderness advocates, and others throughout Alaska and North America. In May the ACS *Bulletin* was reprinted in its entirety as a special issue of the *Sierra Club Bulletin*, with the strategic addition of several glossy photographs of the Alaskan wilderness.

Barry Commoner, a noted plant physiologist at Washington University in St. Louis, was among the scientists placed on the ACS mailing list. Commoner was active with both the Committee on Science in the Promotion of Human Welfare of the American Association for the Advancement of Science and the

Greater St. Louis Citizens' Committee for Nuclear Information (CNI). CNI was an organization created by Washington University scientists and medical doctors that sought to interpret all relevant information about nuclear risks and hazards and to publish this information in terms understandable to nonscientific audiences so that the public might make informed choices about nuclear policies on the basis of accurate scientific information. As Commoner reflected in a 2004 interview, while the "information movement" was born in work fighting the AEC, it succeeded not simply by opposing particular AEC policies but by challenging, on the basis of hard science, the validity of the information on which government (and corporate) policies were based, and by encouraging debates and aggressively demanding answers to the questions that emerged from them.[73] Although the dissident Chariot scientists were geographically removed from debates over radiation safety and nuclear testing, they were aware of Commoner's burgeoning scientific information movement and saw that it matched well with their own oppositional scientific work. Through CNI, the concerns that had been silenced in the reports of Wolfe's committee might be voiced or, better yet, evaluated, through what had become a respected and scientifically legitimate activist organization.[74]

After reading the ACS *Bulletin* and corresponding further with Foote and Pruitt, Commoner and his colleagues in the CNI decided to devote a special edition of the organization's newsletter, *Nuclear Information,* to a "complete report on the probable gains and risks" of Chariot.[75] CNI's twenty-page report, published in June 1961, included pieces by Foote, Pruitt, and Viereck describing, respectively, the people, animals, and plants of Cape Thompson, and it reprinted statements on Plowshare by Livermore's Gerald Johnson and the AEC's John Kelly. The report also included two special reviews of the nuclear harbor project: one by Washington University physicist Michael Friedlander, evaluating Livermore's fallout predictions, and one by Commoner assessing Chariot in terms of "biological risks." While the articles by Foote, Pruitt, and Viereck—as dissenting participants in the AEC's bioenvironmental studies program—lent a certain currency to CNI's report, the reviews by Friedlander and Commoner provided a crucial external evaluation of the oppositional scientists' claims, serving to counter the AEC committee's claims to expertise in its evaluation of its contract scientists' reports.

Friedlander examined Livermore's available published reports on Chariot and the nuclear earthmoving enterprise. He argued that since fallout predictions were based on only the four small (non-Plowshare) cratering experiments in Nevada, Livermore expectations of 5 percent fallout release from the explosives could in fact be grossly underpredicted. For Friedlander, the theoretical curve on which Livermore had based the relationship between scaled depth of burial and vented radioactivity yield could have been constructed quite differently, so that "the yield could be five times the predicted 5%."[76] This uncertainty, along with difficulties in predicting wind conditions around Ogotoruk Creek, suggested that Sr-90 predictions for the Chariot fallout sector up to twenty-eight miles downwind from the blast could be elevated anywhere from three to as much as thirty *times* their present level. Commoner, in turn, used these parameters to inform his assessment of biological risks. Because of stratospheric conditions, Commoner explained, relatively low amounts of worldwide fallout from previous tests had accumulated in polar regions, and so the level of Sr-90 on the ground in northern Alaska was in fact quite low.[77] And yet measurements of Sr-90 and other radioisotopes in Alaskans' bones already indicated somewhat higher strontium counts than for people in temperate zones. Tipped off by Pruitt's questioning of high Sr-90 levels in lichens and in caribou stomachs and antlers, Commoner made a comparative analysis, from the scant data available, of food webs and plant physiology from other northern latitude ecosystems. His best explanation, based in part on comparison with reindeer studies from Norway (and explicitly refuting AEC claims about the relevance of data on Nevada cattle to Alaska caribou), was that caribou had an affinity for food rich in Sr-90. Lichen, an unusual plant which absorbs its water and salts from the atmosphere instead of from the ground, was a major food source for caribou in Alaska. Because the radioactivity taken up in the lichen had not been filtered through the soil, it likely accounted for the high levels of radioactivity in the caribou.

When these ecological dynamics were put together with other data from the environmental studies, including, on the one hand, the observed high winds at Ogotoruk Creek that kept one-third of the region free of snow even in the early spring, and on the other hand, Foote's data signaling the importance of caribou as a food source for the Point Hope Eskimos, Commoner felt confident in drawing three conclusions, aimed not at deciding whether or not the project was definitively safe or unsafe, but rather at how best to characterize the uncertainties associated with the evaluation of Chariot's biological risks:

- Because of the unique mineral nutrition of the lichens, which avidly absorb fallout, the caribou—which feed on lichens—and the humans who eat the caribou, appear to be higher in Sr90 content than any other group in the world.
- The fallout from the proposed Chariot explosion will add to the present Sr90 levels by an amount that cannot now be estimated with any degree of precision.
- In the present state of knowledge about the effects of radiation, no firm prediction can be made regarding the ultimate harm that may

> result from either the present Sr90 levels, or from any increase that may be brought about by Chariot fallout. According to the current philosophy of radiation protection, however, it is assumed that every increase in radiation exposure carries with it an increased risk of disease.

Commoner added that his examination had looked at only a single food chain, but that Sr-90 might also be taken up by the Eskimos through other sources. He listed nine further studies of Sr-90 absorption in Eskimos and their environment that were needed to reduce the uncertainties that stood "in the way of a decision on the wisdom of setting off the Chariot explosion."[78]

With both the CNI report and the committee's official First Summary Report released to the public in early June, the budding scientific controversy over Chariot—much as the oppositional scientists had hoped—attracted immediate national media attention. The *New York Times* covered the release of the *Nuclear Information* issue with an article highlighting the lichen-caribou-human food chain as discussed by Commoner.[79] Less than a week later the *Times* covered the release of the AEC report by quoting the committee's statement that "it would appear that radiation effects would be negligible, undetectable, or possibly non-existent in areas distant from the excavation."[80] The same article also referred to the dissenting report from CNI, however, noting that the AEC "did not deal with one of the key points considered in the St. Louis group's study. This was the particular affinity of lichens, which gain their nourishment from dust in the air, to absorb fall-out material such as strontium-90. Since caribou feed on lichens and Eskimos on caribou, the St. Louis group pointed to the danger that fall-out from the explosion could contaminate the food chain leading from plants to humans."[81] Three months after Wolfe and his colleagues had excluded Pruitt's far less categorical references to lichens from the First Summary Report, they thus found Commoner's ecology lesson turning up in the *New York Times* and other papers. Wolfe's committee was now irrevocably connected to CNI in the popular media, and "controversy"—competing claims to truth based on different evaluations of essentially the same environmental studies—was now stamped on Chariot for all to see.

The next public arbiter was the journal *Science*, which addressed the controversy in its "Science in the News" section with an article titled: "Project Chariot: Two Groups of Scientists Issue 'Objective' but Conflicting Reports." For the editors of *Science*, as the ironic use of quotation marks hinted, "the sharp contrast between the reports suggests that one or both must fall considerably short of the objectivity intended by the authors of each."[82] The AEC report was criticized for providing no evidence for its conclusion of negligible radiation effects, no definition of what was meant by areas "distant" from the throwout zone, and for containing nothing at all regarding the absorption of Sr-90 in the regional food chain. But perhaps sharper criticism was reserved for the CNI report; the editor argued that *Nuclear Information* was not informing so much as misleading the public in its claims that the ultimate harm from Chariot fallout could not be predicted, with Friedlander's methods drawing special criticism.[83] But such criticism only made the controversy grow larger; Commoner and Friedlander submitted a lengthy letter of rebuttal to the editor, contesting *Science*'s critique line by line, which allowed the CNI scientists to forcefully restate their concerns about Chariot to the readership of *Science*.[84]

CHARIOT MISUNDERSTOOD?

Internally, the AEC responded with an effort to identify weaknesses in CNI's analysis so that, publicly, it could set the record straight. From the AEC perspective, CNI had "based its criticisms on a limited view of the Commission's data and experience."[85] Yet of course Friedlander and Commoner had based their criticisms on all publicly available data, along with the help of Foote, Viereck, and Pruitt, scientists who had found it necessary to move outside the AEC's normal channels of communication as a means of making their own concerns public. What had CNI scientists misunderstood in its review of the project?

Stung by the criticism, John Wolfe was particularly upset that CNI had called its special issue a "complete report" when in fact no mention was made of the dozens of other environmental studies being carried out under the committee's direction (see table 2.1). Instead, in its more narrow assessment of risk from radioactivity, Wolfe argued, the CNI report "seemed to anticipate findings for which the investigators have not yet completed their analysis of data. This anticipation of possible results we deem to be unwarranted and certain to arouse excitement before the scientifc picture is complete."[86] On its own, this argument might have been a reasonable one. But given the "anticipation of possible results" already set forth in his own committee's report, and given his own censorship of several investigators' findings, Wolfe's protests appear disingenuous. Another error in the CNI report, as Wolfe argued, was that, like so much of the dissent over Chariot, it had failed to grasp the primary scientific value of the environmental studies:

> One of the major objectives of the bioenvironmental program from its inception, and woven into the plan of every feasible project, was to establish baseline studies. These baselines, biological and physical, were to be the starting points from which bioenvironmental changes, if any, could be measured or observed *after* nuclear detonation. Nowhere in the free world at least, and probably no where [*sic*] else, has such an

> assemblage of bioenvironmental researches been carried on simultaneously in the same location encompassing both land and sea environments. That this project will have scientific potential for decades to come, whether or not there are detonations, seems self-evident, and if there is an excavation by nuclear devices, predictions can be tested and questions answered that can be achieved by no other means. It is therefore of some interest to us as to why this aspect of the long view was ignored in the complete report.[87]

This experimentalist view of the environmental studies was decidedly different from the kind of evaluation that CNI considered essential. For CNI, questions of Chariot's safety were not to be "answered affirmatively before they could even be raised" simply because the experiment would generate new ecological knowledge, of a kind, that could be applied to the evaluation of future nuclear earthmoving projects.[88]

Others on the Plowshare staff found CNI's analysis similarly deficient, revealing what they saw as a profound ignorance as to how experimental science worked: indeed, "if both the vented radioactivity and crater dimensions could be predicted with certainty there would be no justification for the Chariot experiment."[89] But Wolfe, his hackles raised in the candid, internal AEC circular, was unwilling to concede even this basic element of uncertainty. He was irritated that the CNI report was devoted almost entirely to Chariot's potential fallout hazard. "What fallout hazard?" he asked.[90]

In view of the repeated downward revisions of "permissible doses" of radiation from the Federal Radiation Council during the 1950s (see chapter 1), and bearing in mind the ongoing, unsettled controversies over genetic damage and increased cancer and leukemia risks from low-level radiation exposure, Wolfe's confidence that fallout from Chariot would not at least potentially pose a serious health hazard remains difficult to reconcile.[91] As head of the environmental program, he was presumably aware that, according to a 1961 letter from Gerald Johnson, Livermore's Plowshare director, to the AEC's John Philip, on the basis of an optimistic prediction of 5 percent release of radioactivity, 1.5 billion curies of fission products (as measured one hour after the blast) were expected to vent from the 280-kiloton Chariot configuration.[92] As planning continued for a possible spring 1962 blast, Livermore test managers and AEC shot engineers actually expressed more concern for the consequences of Chariot fallout than did the DBM ecologist Wolfe.

New data on surface winds and wind shears at ground zero became available in the spring of 1961, and with them new problems arose in terms of practical, operational planning for Chariot, the domain of the AEC's Nevada Operations Office. As James Reeves, the assistant manager for test operations in

Nevada, reported to the AEC's John Kelly, these new data indicated that it would "be rather difficult to select without protracted waiting periods a firing date in which the fallout would be entirely confined to the desired inland sector."[93] But a seaward shot presented additional risks, including the possibility of sending fallout across the Soviet border, just 180 miles away, and in the event of an unexpected change of wind direction, of sending close-in fallout across the nearby villages of Point Hope and Kivalina. Although AEC planners may not have shared CNI's concern for the "special biological situation" of the Cape Thompson area, they could hardly disagree, with two villages located within forty miles of ground zero, with Commoner's assessment that there was "little margin for error."[94] And yet Livermore and AEC scientists and engineers never seem to have doubted that they could operate safely within these margins—perhaps it was this confidence that Wolfe had tapped into, secondhand. As Reeves rationalized for the AEC's San Francisco Operations Office, more than one hundred nuclear explosives tests had already been conducted in Nevada, which was a more densely populated area than the Cape Thompson region. Since, by Reeves's estimate, it was "apparent that Chariot devices could be detonated safely and with relative ease at NTS, even considering the 'regulation' limit . . . it is reasonable to conclude that it would be much easier to accomplish the event safely at the Chariot site."[95] In fact, Chariot was to have a yield almost four times greater than the largest test to date in Nevada, and even much smaller Nevada tests had not been without controversy and significant radiation exposures for downwind populations.

While AEC operations planners continued to grapple with the constraints placed on them by the physical geography of Cape Thompson, the mailboxes of John Kelly and other AEC and Livermore personnel began to fill up with questions from politicians, scientists, and concerned citizens and activists. The scientific controversy that had emerged was focused on the biological consequences of Chariot, with a special concern for the Eskimos that depended on the environment for their livelihood. Yet little attention had been focused, by either the AEC or its scientific opponents, on the juridical question of whether the AEC actually had the *right* to blast a harbor (or, more accurately, a hole in the ground) at the mouth of Ogotoruk Creek. Who held legal rights to the land, what was the nature of those rights, and under what authority had they been granted?

PROPRIETARY CLAIMS

When John Kelly, in the midst of these scientific controversies, visited Alaska with operations specialist Reeves, public relations man Rod Southwick, and AEC commissioner Leland Haworth in July 1961, he was pleased to find state legislators in Anchorage and Fairbanks seemingly unopposed to Chariot. Although some felt that more effort should be made to assure the people of Point Hope and Kivalina of the safety of the project, others, Kelly reported, suggested that the AEC "consider going back to the original larger experiment to make a full size harbor to aid in the recovery of copper from deposits alleged to be found in the Ogotoruk vicinity."[96]

The AEC entourage arrived in Kivalina on July 29. As Kelly observed, the "Kivalina Village Council did not appear to have any strong feelings one way or the other about Chariot." The same day, however, the group went to Point Hope, where they met with an audience of between 125 and 150 people. The meeting lasted for over four hours. Kelly reported, "The Point Hope people seemed to be genuinely concerned about the experiment. Most of this concern was based on a lack of understanding of the project, plus some distrust of the methods employed in pursuing the site selection work which has been underway. There is a difficult communications problem in talking with these people." This communications problem—a problem that Kelly had not noticed among the seemingly disinterested council members in Kivalina—was largely a product of language barriers but also stemmed, as Kelly understood it, from a generation gap among the Eskimos. There were "quite a number of penetrating and well-thought-out questions that came primarily from the younger Eskimos and from the Lawtons [the Episcopal missionary and his wife]," but Kelly complained that "many of the questions and our answers had to be interpreted by members of the Village Council into the Eskimo tongue. The older people appeared to be somewhat emotional and obviously did not understand the project nor our assurances that it would only be conducted if safety could be assured."[97] For many at Point Hope, however, emotion was not so easily disconnected from what they understood as fundamental questions of health, livelihood, and sovereignty.

Don Foote was not in the village at the time of the meeting, but he had not left the region. Although his contract expired in the spring of 1961, not to be renewed by the AEC, he had elected to stay at Point Hope for another year without AEC support to continue research on both historical and contemporary Eskimo land-use patterns in Point Hope and Noatak and, judging by his actions, to stay involved in anti-Chariot politics as well.

For his part, Kelly felt that the Alaskan trip had been a good one. He surmised that the visit to the Eskimo villages had done "quite a bit to allay the fears" about Chariot, and he postulated that, at Point Hope, "post-visit discussions from among younger Eskimos will help to allay some of the fears of the older people."[98] By the time his public relations team had reached Point Hope, however, the village had already begun to organize its opposition and to take it above the heads of the AEC to other federal offices. In March 1961, for example, the women of the Point Hope Village Health Council submitted a letter to President John F. Kennedy stating that they were "concerned about the health of our children and the mothers-to-be after the explosion."[99] But while health hazards from fallout may still have been seen largely as matters for medical scientists and geneticists to decide, the politics of Point Hope's opposition to Chariot began to turn decidedly in the village's favor when they shifted the focus of their protests to the issue of land rights. Here the language barrier that Kelly had perceived did little to obstruct Point Hope activists as they forged new connections with outside actors and institutions.

Among the most important of these connections was with the New York–based Association on American Indian Affairs, which offered legal assistance to Point Hope in its struggle.[100] Aided as well by additional research from Foote, Les Viereck and his wife, Teri, and others looking into the AEC's original land withdrawal from the Bureau of Land Management (BLM), the Point Hope Village Council challenged the legitimacy of the AEC's land claim from the public domain. Although the BLM had researched mineral rights to the 1,600 square miles withdrawn for Chariot, it had not examined extant native claims to the land. And although Point Hope did not hold property rights (title) over the land per se, the Organic Act of 1884, codified shortly after Alaska was purchased from Russia and reaffirmed in the Alaskan Statehood Act of 1958, gave the village *proprietary* rights "to the *use* of the lands 'now claimed by them' that 'shall not be disturbed' in the use thereof."[101] The legal argument was ultimately set out in a July 1961 letter from the village council to Interior Secretary Stewart Udall, the former Arizona congressman who had just months earlier been appointed to the Cabinet position. After explaining to Udall the significance of the Cape Thompson region to Point Hope's subsistence—claims that were backed up by the submission of Foote's interim reports and data—the letter raised a question that neither AEC bureaucrats nor Livermore physicists could claim the authority to answer: "Why was the Bureau of Land Management, which we understand is under the Department of the Interior, given any right to allow the Cape Thompson area of land to be utilized by the Atomic Energy Commission for the explosion of Project Chariot? . . . We, the people of Point Hope, have a claim to this land and we consider your office, the Department of Interior, to be the protector of our rights."[102]

Udall was "touched profoundly" by the letter, according to his assistant. Having grown up amid Navajo and Hopi communities in rural northern Arizona, Udall was not unfamiliar with native land rights issues, and as a Democratic congressman he had recently served on a Joint Committee on Navajo-Hopi administration. He would go on to earn a reputation, as secretary of the interior for both Kennedy and Lyndon Baines Johnson, as an advocate for environmental conservation as well as for Native Americans. The Interior Department responded quickly to Point Hope's letter by exploring the question of Eskimo land rights in greater detail. Meanwhile the dissident scientists, with the help of the ACS *Bulletin* and CNI report, kept Udall apprised of the scientific controversies and ethical misdeeds associated with the Chariot bioenvironmental program. By August the Interior Department had clarified not only that Point Hope held legitimate proprietary rights to the land, thus affirming the Eskimos' rights to hunting without interruption in the proposed landward fallout sector, but also that the AEC did not have an official "withdrawal" on the land, but rather a "use permit" from the BLM. Although the AEC had been permitted to conduct its preliminary studies, including the environmental studies along with a series of chemical high-explosives tests at the project site, it had been waiting to pursue the actual land withdrawal until permission for the nuclear explosions had been granted by the president. As secretary of the interior, Udall was in a position to block the land withdrawal—if he was not satisfied with the AEC's assurances of the Eskimos' safety—before the instant harbor could be blasted.[103]

From Cape Thompson to Area 10

The Soviet Union resumed nuclear testing on September 1, 1961, signaling an end to the voluntary test moratorium among the United States, United Kingdom, and the Soviet Union that had been in place for nearly three years. The United States followed just two weeks later with a series of underground tests in Nevada. Along with the return to weapons testing, this not wholly unanticipated turn of events meant that Plowshare's experimental program could begin at last, without fear of being stigmatized for breaking the moratorium. As operational planning for Chariot continued, Livermore scientists quickly readied another Plowshare experiment, with little controversy, 1,200 feet below ground in a salt formation southeast of Carlsbad, New Mexico.

Project Gnome, as it was called, was an experiment designed for a number of purposes, including the recovery of usable power from the heat generated by a nuclear explosion, the recovery of scientifically and commercially valuable isotopes, and a range of additional experiments exploring aspects of neutron physics, nuclear-device design principles, and nuclear explosive effects in a salt formation. The 5-kiloton explosive was detonated on December 10, 1961, and later described for Congress as an "extremely successful experiment." In fact, Gnome had provided fairly conclusive proof that power recovery was *not* viable from underground nuclear explosions at present, mainly because the heat from the blast destroyed much of the infrastructure designed to contain it. Another surprise from Gnome was the discharge of highly radioactive vapor that "began issuing from the shaft [drilled for placement of the explosive] shortly after the detonation," and "continued through the next day." The public safety program, which included blocking off nearby Highway 128 for several hours and washing down the people from seven vehicles that had been on the highway before the roadblock (along with some of the on-site technicians), was also confirmed as a success by the AEC. Despite these accomplishments, the AEC was compelled to admit, in its annual report to Congress, that the "exact mechanism by which this escape of radioactivity occurred is not yet known."[104]

Unfortunately for Plowshare's advocates, Gnome had been set up as a kind of international spectacle to show off America's peaceful atom in action, with over five hundred observers from sixteen countries in attendance just a few miles from ground zero. But with the unexpected venting and, as a front-page *New York Times* article reported, an escaped radioactive vapor cloud on the loose, the AEC had been left with a highly public embarrassment, even as it insisted that there was no danger from the vented radioactivity.[105] And however successful the AEC could call Livermore's Gnome experiment, the test could have done little to instill public confidence—or that of President Kennedy—in Livermore and AEC safety predictions for the much larger and more controversial Chariot blasts.

Undeterred, the AEC put Chariot's best face forward to Congress in January 1962, claiming that "the criticism of Project Chariot, apparently emanating from a few small groups . . . appears to have substantially abated."[106] It is notable that although the opposition had given no sign of acquiescence, AEC public information writers felt compelled to make such a claim publicly in an annual report. Such delusions would prove to be short lived. In late January Udall informed AEC chairman Glenn Seaborg that the Interior Department, in light of its role in the land withdrawal process, would like to participate in the AEC's review of the Chariot environmental studies when they were complete.

Making matters worse for Plowshare, in April an article appeared in *Harper's Magazine*, coauthored by wilderness writer Paul Brooks and Don Foote's brother, Joe, entitled "The Disturbing Story of Project Chariot." The piece was another conduit for Don Foote's human geographical and ecological studies, and for his makeshift public history of the project as well.[107] While the CNI report had sought to maintain a properly objective and scientific attitude, the *Harper's* article was written quite explicitly to mobilize the public against Chariot. Given the frequent changes in Chariot's public posture and its experimental design, the dubious claims of Wolfe's Committee on Environmental Studies, and not least, the protests of Eskimos and dissident scientists alike, the case against Chariot was by this time not a difficult one to make. In explicit counterpoint to the heroically modern rhetoric that Edward Teller and other Plowshare enthusiasts had used to promote Chariot, Brooks and Foote concluded their article by turning their gaze on Teller himself:

> "If your mountain is not in the right place," said Dr. Teller at Anchorage, "just drop us a card." It was only half a joke. Our ability to alter the earth we live on is already appalling. Few of us are in a position to judge the ultimate scientific value of an experiment like the *Chariot* explosion. But it *is* up to us to know what is going on in that far corner of the United States. And to realize that another scale of values is also involved: not the precise relations between depth of burst and crater characteristics, but the precise relations between unlimited power and the awesome responsibility that goes with its use.[108]

Public reaction to the *Harper's* article was swift, and the AEC and Livermore again found themselves justifying the experiment to politicians and concerned citizens across the country.[109]

By this time, however, Plowshare scientists were already considering a way out of Chariot. In March the AEC's Kelly had asked for Livermore's "recommendation" on "what action the laboratory believes should be taken with respect to Project Chariot."[110] In light of the scientific and legal controversies then surrounding Chariot, approval from President Kennedy, let alone Secretary Udall, must have seemed a long shot at best. The laboratory concluded that Chariot was no longer justified on technical grounds, and recommended that the project be canceled. Yet new Livermore director John Foster was concerned that the cancellation "could have repercussions which would adversely affect the whole Plowshare program."[111] The cancellation of Chariot might imply that Livermore agreed that nuclear excavation was unsafe or, perhaps worse, that the AEC had backed down in the face of resistance from a few hundred Eskimos, some committed biologists and conservationists, and a geographer who had flunked his Ph.D. comprehensive exams.

Unable to detonate in Alaska, Livermore decided on a retreat to the privacy of the AEC's Nevada proving grounds—where more than a decade of nuclear testing, primarily in the atmosphere, was testament to its authority—to try the first of Plowshare's nuclear excavation experiments. Livermore recommended to the AEC that Chariot be canceled on April 30, 1962; on May 8, the National Security Council authorized its replacement, Project Sedan, a single-shot 100-kiloton nuclear cratering test to be conducted in the desert alluvium of Area 10 at the Nevada Test Site on July 6, less than two months away.[112] When this choice was made, John Kelly, reflecting the concerns of

Livermore's Foster, was worried that Chariot's "cancellation will contribute to the skepticism on the safety of nuclear excavation." Kelly proposed a public information plan designed explicitly to counter such suspicions: "The public announcement of the decision to cancel Project Chariot should be carefully timed to coincide with a convincing event which documents the technical rationale for the decision. Project Sedan will be such an event. Therefore, it is proposed to include, in the second post-shot announcement of the results of Project Sedan, the announcement of the cancellation of Project Chariot."[113] Still, as Kelly understood it, while some of Chariot's scientific-technical objectives could be met by Sedan, other important project goals would not.

In drafting his letter justifying the cancellation of Chariot, Livermore's Foster had listed five objectives for the Alaska experiment:

1. Extension of nuclear cratering knowledge to a different medium, and to higher yields including the investigation of air blast and seismic shock.
2. Determination and distribution of radioactivity.
3. Investigation of the effects of simultaneous multiple detonations.
4. Determination of the effect of action of water on a large crater.
5. Effects of a large scale nuclear detonation on the biosphere.[114]

Sedan would indeed satisfy some of these goals, perhaps even better than Chariot would have. From the beginning, the primary technical purpose of Chariot had been to work out scaling laws in the 100-kiloton range for depth of burial as it related to crater size and radioactivity release. In Sedan, by detonating only a single explosive, the morphological characteristics such as crater size, slope stability (of the crater produced), and "throwout" of earth and rock could be measured more precisely relative to the phenomenology of the nuclear explosion. But Sedan would not extend cratering knowledge to a different soil medium; it would not provide data on simultaneous "row-charge" cratering explosions; it would not determine the effects of water action on a large crater; and, though the distribution of radioactivity would be monitored, without extensive preshot bioenvironmental baseline studies—the Test Site would have made for a highly unusual "baseline"—there would be little useful information on the effects of large-scale nuclear excavation on the biosphere.

But there was also more to Chariot than the fulfillment of these scientific-technical experimental objectives. As Foster put it candidly in an early, unsent draft of the letter canceling Chariot: "Perhaps never stated, but of course always kept in mind, is another purpose of Chariot, namely: . . . To dramatically demonstrate (a) the capability of nuclear explosives to excavate large volumes of earth in a controlled and constructive manner, and (b) that such excavations can be done safely with minimal hazard."[115] Still, Foster remained optimistic that, in the future, Livermore could develop an economically useful demonstration project, and not just a "model harbor."[116] In the wake of the public scrutiny unleashed in Alaska, however, the AEC's Kelly, for one, was keenly aware of how the politics of place had bolstered the resistance to Chariot. As Kelly understood it, an "experiment comparable to Chariot" was "essential to the excavation program," and yet "the selection and safety evaluation of an alternate site will be very difficult."[117] Meanwhile, as planning for the Sedan test proceeded rapidly, no one from Point Hope or Fairbanks was informed of Chariot's cancellation.

Geographies of Authority, Rights to Knowledge

In their urgent desire to better understand the phenomena of nuclear cratering and its effects on the biosphere, *and* to dramatically demonstrate their geographical engineering techniques, Plowshare scientists hoped both to "try" and to "show" their experiment in Alaska. Yet for Project Chariot, the problem of creating order within which dissent could flourish—the production of an experimental space—was made to turn on boundary issues. To demonstrate nuclear excavation as a viable scientific technique, program scientists sought to create this new experimental space outside of their own laboratories and proving grounds. But by moving to more public spaces, they found, the boundaries of scientific authority and dissent also shifted. Livermore's authority was challenged most effectively by those whose authority rested more firmly on the place itself or on knowledge of the place: the Eskimos from Point Hope, who didn't see why a "model harbor" was needed on their caribou hunting and crowbill egg–gathering grounds, and University of Alaska and other polar regions scientists.

Writing to his brother Joe in the spring of 1961, and venting frustrations accumulated in the difficult work of organizing opposition to Chariot, Don Foote reflected on the interplay of authority, rights, and social responsibility:

> My sadness stems primarily from the endless discussions I have had wherein the idea that individuals do not have the "right" to seek truth beyond the limits of [their] specific field were presented as reasons against my actions or as reasons for others [*sic*] inaction. What in-fact these individuals were proposing is, of course, that scholars have no "moral right" to speak on those subjects in which they do not have a phd. But when you point out the hypocritical position these people take when they accept as almost god-given truth the words of Edward

> Teller on nuclear physics, genetics, international politics, philosophy, and radiation biology, they become highly irrational and the communication ceases. They *want to believe* an authority. *They want to give up their responsibility.*[118]

Questions of scientific authority, as Don Foote had come to understand them, were at times inseparable from those of politics and morality. While scientific truth remained, for Foote, an ideal or principle to be approached through the normative rules and rational procedures of scientific observation, analysis, and interpretation, these procedures could only work so long as scientists acted morally; then science was moral. Foote's ongoing appraisal, throughout the Chariot studies, of the honesty and sincerity of the scientists involved underscores this point. Foote was particularly concerned that scientific authority and the social responsibility that came with it had been fractured, constrained by the boundaries of disciplines and expertise. But this was only the case for some scientists. Others, like Teller and his colleagues, were able to transcend such boundaries, using them to their advantage much as they had mobilized the boundaries of classification in the effort to control the nature and extent of public information in connection with the risks posed by nuclear testing. And yet where Teller had spoken of atomic miracles and scientific progress as things universal, like the iron laws of physics, the story of Chariot suggests possibilities for understanding science in practice as a much more varied spatial process, carried out, in different ways, by scientists, engineers, public information specialists, journalists, and Eskimos who weighed in their harvests of seal and caribou and sought partnerships to help them prevent the project.

In a down moment, it had seemed to Foote that AEC authority, whether in the personage of Teller, Wolfe, or Southwick, was as universal as its claims to the benefits of geographical engineering. Yet when this authority was effectively challenged—challenged, as Foote phrased the problem precisely, by oppositional scientists, Point Hope Eskimos, and other activists claiming and asserting their own rights to knowledge (that is, "the 'right' to seek truth beyond the limits of [their] specific field"; the "'moral right' to speak on those subjects in which they do not have a phd")—the legitimacy of many of the AEC's claims was effectively undermined, and Plowshare was forced to retreat to the more controlled spaces of the Nevada proving grounds. What was safe or feasible in one place, the Plowshare scientists' strategic reactions to the Chariot opposition and controversy made clear, was not necessarily so in another. What is more, the affirmation of these rights to knowledge also encouraged demands for additional rights and the production of new political movements, including the Alaskan native land rights movement that gathered significant momentum—and garnered key allies—in the resistance to Chariot.[119]

Just before ten o'clock on the morning of July 6, 1962, Livermore project scientists, together with AEC Nevada Test Site scientists, engineers, and technicians, counted down to the detonation of Sedan, Chariot's replacement experiment, from a bunkered control center on the grounds of the Nevada Test Site. As the 100-kiloton blast erupted beneath the desert floor of Area 10, just ten miles from the northern border of the Test Site, it soon became clear that, despite its tactical retreat from Alaska, Plowshare would not be easily confined within AEC territories.

Chapter 4 Nuclear Craters

THE BOMB SAT beneath 635 feet of soft desert alluvium and a clear Nevada sky on the morning of July 6, 1962. Although the rapid, two-month planning period for Project Sedan had, according to its Livermore project managers, precluded "many pertinent experiments [that] could have been designed and included," there had still been time enough to fit the Nevada Test Site with an experimental infrastructure for gauging many aspects of the explosion and its effects. Six photographic stations were set up at contrasting angles and distances to observe surface motion and dust-cloud dimensions. An assortment of tarpaulins, trays, and measuring rods were placed at various points of orbit around ground zero to measure "throwout" of earth, rock, and dust. Air blast and seismic monitoring instruments were situated both on- and off-site, and biological sampling plots had been quickly established to study at least some biological effects of radiation from Sedan. Fallout trays were distributed throughout the planned fallout sector to sample radioactivity release. Meanwhile, two pens of live beagles were placed at stations thirty-one and forty-two miles from the bomb as part of an experiment on radiation measurement techniques, contracted through the U.S. Public Health Service, which would observe the accumulation of inhaled radioactive iodine in the thyroid gland, and there the dogs waited as project scientists counted down to the detonation of the first nuclear cratering experiment to be carried out under the Plowshare program.[1]

At ten o'clock, the bomb was detonated—the largest nuclear explosion ever in North America at that time, with a yield later calculated at 104 kilotons—setting off a rumbling base surge that extended outward from ground zero in a wavelike motion in the earth's surface. Within three seconds, a spherical dome of earth six hundred to eight hundred feet in diameter had risen to a height of nearly three hundred feet above the ground, followed by a bright flash as the explosion fractured the landscape's surface (figure 4.1). The dust cloud that the blast had produced then climbed to an altitude of twelve thousand feet, where it was topped by inversion conditions in the atmosphere and distributed downwind. It left behind a crater 1,200 feet wide and 320 feet deep, a volume remarkably close to the geographical engineers' predicted parameters.[2] Many results, however, particularly those associated with radiation, would turn out in the hours, weeks, months, and years of analysis ahead for Sedan to be much further from the predictions of Plowshare scientists. The dust cloud was 50 percent larger than predicted, and five times more fallout than predicted had come down in the local fallout pattern.[3] And yet, even after early public pronouncements (seemingly based on predictions) estimating that 95 percent of the bomb's radioactivity had been trapped underground were later downgraded to just over 90 percent, the experiment continued to be declared a success by Plowshare supporters. As the AEC announced in a July 7 press release, the 7.5 million cubic yards of dirt and rock ejected by the bomb—and the great hole punched in the desert floor—signaled a "significant contribution to nuclear earth-moving technology."[4]

By the following spring the AEC judged that radiation hazards around the crater had lessened enough to allow a photographer from *Life* magazine to explore the crater. The story that followed in the June 1963 *Life*, beneath the heading "H-Blasted Crater Twelve Million Tons Deep Makes a Lab for AEC Men," expressed the triumph of Sedan more boldly than had earlier AEC press releases. With three pages of photographs and a three-paragraph article, the magazine showed how the science of nuclear earthmoving was now an experiment-in-progress inside what had been named Sedan Crater (figure 4.2). At the rim of the quarter-mile-wide crater, an Army Corps of Engineers team was at work sinking a drill into the rim of the crater to determine slope stability, while at the bottom of the steep, 320-foot-deep cavity, Livermore scientists and technicians were beginning to examine the underground layers of earth, expecting to find, *Life* reported, hundreds of feet of "fallback soil" and "a few feet of glassy material" that would tell them a great deal about the physical phenomenology of nuclear cratering explosions.[5] The grounds of knowledge rested on knowledge of the grounds.

At least one other prediction proved to be accurate: John Kelly's preordained public information plan indicating that the technical accomplishments of Sedan would provide occasion for announcing the cancellation of Chariot, the controversial experiment that Sedan had been designed to replace, or at least Chariot's de facto cancellation. In an August 24, 1962, press release heralding Sedan's successes, the AEC announced that a decision on Chariot had been "held in abeyance." Although it insisted that the Alaskan "bioenvironmental

Figure 4.1. Time-series photographs of Project Sedan nuclear-cratering experiment, July 6, 1962, Nevada Test Site, at (reading down) 1.9, 2.8, 4.0, 27, 39 seconds, and 6 minutes after detonation. The bar in lower left corner represents 300 meters. *Credit must be given to the University of California, Lawrence Livermore National Laboratory, and the Department of Energy (Atomic Energy Commission) under whose auspices the work was performed when this information or a reproduction of it is used.*

studies . . . have not revealed any way in which Chariot would endanger the local inhabitants or their livelihood," the AEC now argued that "some of the data originally planned to be obtained from Chariot are now available or may be developed from other experiments."[6] Legally, the AEC had merely deferred its recommendation to conduct the experiment—Chariot had not been canceled per se, leaving an abiding uncertainty in the minds of Eskimos and other

Figure 4.2. Sedan Crater, Nevada Test Site. Trucks and equipment at the lower right side of the crater provide a sense of scale. *Credit must be given to the University of California, Lawrence Livermore National Laboratory, and the Department of Energy (Atomic Energy Commission) under whose auspices the work was performed when this information or a reproduction of it is used.*

anti-Chariot activists in Point Hope and Fairbanks—but the campsite at Ogotoruk Creek was placed on caretaker status and the experiment dropped off the map of Plowshare planning.

As we saw in chapter 3, the move from Chariot to Sedan offered a geographical solution to the problem of opposition to Chariot. But the solution presented new epistemological problems for the excavation program: how could the knowledge produced in Sedan and in subsequent cratering tests and studies be authenticated, even though the tests, and much of the data produced in them, were carried out essentially outside of the public view? And how could new knowledge, derived from the coded areas of the Test Site's artificial lunar landscape, be made to apply to different geographies and ecologies?

Life's celebration of the ingenuity and expertise of the "AEC men" in turning the massive crater into a "lab" made no reference to Chariot. Nor did it mention the fallout from Project Sedan—first radiological, then political—

after the isotope iodine-131 was detected at high levels in the dairies and front lawns of Salt Lake City and much of central and northeastern Utah in the week following the earthmoving blast. Instead *Life* and other supportive periodicals, such as *Business Week* and *U.S. News and World Report,* accentuated only the most persuasive aspects of the experiment.[7] The fact that a thermonuclear explosion had moved twelve million tons of earth—tangible evidence that could be seen in photography depicting tiny Test Site workers at the foot of the "largest excavation ever created by a single man-made explosive"—was privileged over any sign of controversy, thus promoting the experiment as a convincing success.[8]

For some, Sedan *was* a convincing success. It had provided an impressive store of new data for Plowshare scientists to (at last) begin to refine their knowledge of scaling factors and radioactivity distribution from nuclear cratering explosions. For the AEC, after just one day's reflection, it had already shown "that earth-moving projects of this kind can be conducted, under controlled conditions, in safety."[9] It had been clear to project scientists that the amount of radioactivity released by Sedan was far greater than had been expected, but public representations of the experiment remained highly selective. The degree to which Plowshare advocates were able to "carefully phrase and limit the subject" (as an earlier AEC public information directive had put it) was critical not just to public knowledge of Sedan and the Plowshare program but also to the practice of nuclear excavation as an experimental science carried out within the classified spaces of the Nevada Test Site and the Lawrence Radiation Laboratory at Livermore. For despite Plowshare's apparent defeat in Alaska, the emergence of disturbing new arguments about the health hazards of ingested radioiodine, and ostensibly severe limitations imposed by the Limited Test Ban Treaty of 1963, its excavation program actually gained unprecedented funding increases for nuclear cratering tests between 1962 and 1968. And yet the controversies that the AEC hoped to suppress with engineered photo opportunities like the *Life* feature and in glowing reports to Congress were not mere inconveniences for the nuclear earthmoving program. Scientific and political controversies and other "outside" events actively structured the nature of the nuclear excavation program: its science, its politics, and its geography. The grounds of knowledge for geographical engineering *could* be found in nuclear craters in the Nevada desert, but only to the degree that the science of Plowshare could be both conceptually and geographically bounded by its advocates, set apart from the controversial world beyond the borders of the AEC's open air lab in Nevada.

This chapter continues by exploring two connected controversies, one emerging after Sedan that focused on the risks posed by radioiodine from fallout and the other on the ratification of the Limited Test Ban Treaty in 1963, before examining how Plowshare's scientist-planners reacted to these controversies and, to a degree, resolved the problems that they posed, through a number of new strategies. While the following chapter will examine in greater detail the Plowshare program's continuing search for off-site demonstration projects, the last section of this chapter focuses on several Plowshare nuclear cratering tests conducted in Nevada during the middle 1960s, and asks about how the use of particular spaces and boundaries allowed Plowshare scientists to get on with the work, and self-styled progress, of an experimental program in nuclear earthmoving.

Sedan's Radioiodine: Knowledge and Controversy

Planning for Sedan in many ways resembled the organization of other atmospheric nuclear tests in Nevada. The overall experimental design was directed by laboratory project managers (in this case, Livermore's Milo Nordyke and Marvin Williamson), but also depended on operational planning procedures carried out by AEC Nevada personnel, including site selection within the Test Site, and public information and public safety programs. Since Sedan was a cratering experiment (and not really an underground test, except in the sense that the explosion was to start underground), "venting of both solid particulates and gaseous products was expected to occur following the detonation." Radiation and fallout predictions "were based on the assumption that the deposition of solid particulates would produce a geographically extensive fallout pattern and that emersion within the passing nuclear cloud could significantly contribute to external gamma exposure."[10]

One area of planning for Sedan shifted radically prior to the July 6 test date. Initially there was to have been an "open" public information policy for Sedan: media representatives were to be invited to witness the blast in person somewhere on-site; unclassified information was to be widely distributed; and still and motion-picture photography of the blast and crater were to be issued as public information releases.[11] Shortly before the test date, however, a decision made at the "national level" to recast the public information plan for Sedan. Photographs of any kind were now to be withheld from public view, for the time being, and observation was restricted to participating Livermore and AEC Nevada personnel. Of course, the largest explosion ever set off in North America—though it was still almost three times smaller than the final, reduced configuration that had been planned for Chariot—could hardly have been kept secret for long. Press releases were prepared in advance for distribution both before and after the experiment, and in keeping with the norms of nuclear testing in

Nevada, on-site radiation safety work was directed by the AEC's Nevada Operations Office while the U.S. Public Health Service was contracted to monitor radiation off-site. Although they had never carried out a 100-kiloton nuclear earthmoving test before, Sedan's safety planners were confident that they could operate close to, if not completely within, established Federal Radiation Council limits for radiation exposure to employees and off-site populations.[12]

The morning of July 6, Plowshare scientists saw many of their theories about large-scale nuclear cratering verified. The fact that the size and shape of the crater conformed so closely to predicted dimensions showed project scientists that the "scaling factors" that told them how deeply to bury the bombs for maximizing crater size—factors developed from previous small-scale ("military") nuclear cratering tests as well as those conducted with chemical high explosives—could now be applied for blasts up to one megaton, an order of magnitude larger than Sedan, even as it was noted in a later caveat that "empirical scaling is inherently unreliable for extrapolation beyond the region of experience."[13] In addition to these physical cratering effects, Plowshare scientists had also expected Sedan to produce new data on radioactivity from nuclear earthmoving explosions, data that would be useful for predicting and evaluating some of the practical public safety problems to be encountered with nuclear excavation elsewhere. But in this area the observed results of Sedan strayed much further from predictions, with almost a full kiloton of fissionable products blasted into the atmosphere. Even if, as the AEC later claimed, three-fourths of the radioactive fallout had been deposited within two and one-half miles of the crater, that still left 250 tons of radioactive dust and debris to be accounted for beyond this range.[14]

Why had fallout been so badly underpredicted for Sedan? According to the 1965 Final Report on Sedan, the surprise was due chiefly to the moisture content of the soil at Area 10, which turned out to be three to six times higher than the data used for predictions. However successful project scientists had been in predicting crater dimensions, in their ability to predict the quantity of fallout produced by a cratering explosion they could conclude only that "the relationship between depth of burst, yield, and radioactivity escape used to extrapolate the Sedan event must be re-examined."[15]

Long before these diagnostics could be performed, the AEC faced an immediate public health crisis as the unexpectedly large dust cloud drifted over the northern border of the Test Site. The cloud was tracked in twelve-mile-per-hour winds along Nevada State Highway 25, which was later scrubbed—along with the cars traveling it—with pressurized hoses and closed off for several days. Ten people living on area ranches were relocated, while others downwind were told only to stay indoors until the cloud passed—including the people of Ely, Nevada, northeast of the Test Site near the Utah border, where street lights had to be turned on at four o'clock on the July afternoon to counter the shade from Sedan's dust cloud.[16] "The cloud moved northeast during the night as had been predicted," the AEC later noted, "and during Saturday [July 7] was believed to be somewhere in Utah, although it apparently had lost its identity."[17]

The dust cloud lost its identity, however, by dropping fallout across the pasturage, farms, and front lawns of eastern Nevada and central and northeastern Utah. Far more contentious claims about fallout from Sedan quickly emerged, most forcefully from radiation ecologist Robert Pendleton, the director of the University of Utah's Department of Radiological Health. On July 7 Pendleton and twenty students were measuring background radiation in the Wasatch Mountains, twenty miles southeast of Salt Lake City, for a Public Health Service–funded project examining levels of the isotope cesium-137. According to Pendleton, a reddish-brown dust cloud approached from the south across the Salt Lake Valley. When the cloud reached Pendleton and his students, "our instruments went completely nuts and we couldn't measure what we were trying to do."[18] The intensity of external, gamma radiation increased to one hundred times normal background radiation.[19] After this encounter with Sedan in the field, Pendleton became one of several scientists who challenged the AEC's official claims about fallout from Sedan and its descriptions of the nature of the hazard that it posed—and had already posed—to Utahans and others living downwind from nuclear testing. The cloud, meanwhile, was tracked by Weather Bureau climatologists as it drifted over Colorado, Wyoming, Nebraska, and the Dakotas (and probably into Canada), before turning south of Chicago on an eastward route to the Atlantic Ocean.[20]

REVISIONIST SCIENCE: I-131 IN UTAH AND THE HISTORICAL ENVIRONMENT

The dust cloud passed over Robert Pendleton and his students on a Saturday; by the following Thursday, Pendleton and other University of Utah health scientists were collecting samples of milk and leafy vegetables from several farms near Salt Lake City. After recording high levels of the isotope I-131 in the samples, attributable to Sedan as well as to several other nuclear explosives tests also executed in early July 1962, Pendleton recommended to state authorities that the contaminated milk be diverted to powdered milk and cheese, so that the short-living radioiodine—which had a half-life under eight days—would decay to safer levels. Yet despite a 100 percent correlation between irradiated milk and cows who had been grazing on pasturage, rather than stored feed, neither the AEC nor the Public Health Service acted to confirm Pendleton's data, and Utah health authorities delayed action on the regulation of foodstuffs until most of the contaminated milk had already been consumed. In turn, Pendleton expressed his concerns publicly. But as the state health director, Carlyle Thompson, explained to Pendleton, such public statements tended to "complicate the performance of our public responsibility and of our working relationship with industry."[21] Thompson—acting out a pattern of compliance with the AEC in testing and fallout matters that had been characteristic of Utah public and religious officials during the 1950s and 1960s[22]—assured local newspapers that there was no danger posed by the fallout. Several scientists working both inside and outside the AEC, however, took the high levels of radioiodine in the state's milk supply quite seriously.

Harold Knapp, a mathematician working for the Fallout Studies Branch of the AEC's Division of Biology and Medicine in Washington, D.C., was during the summer of 1962, in his own words, "trying to put the finishing touches on a report on *Radiation Exposure in the United States from Nuclear Test Fallout* and discovered at the last minute that the most important problems concerning the most important nuclide hadn't been thought about."[23] For more than a decade of atmospheric nuclear testing during the 1940s and 1950s, external emitters of radiation, especially long-living radionuclides like strontium-90 and cesium-137, were the health hazards of primary concern in connection with atmospheric nuclear testing. It wasn't until the first years of the 1960s that many public health scientists, reacting to renewed testing following the 1958–1961 moratorium and to new knowledge of radioactivity hazards following a British reactor fire in 1957, began to look more seriously at ingested internal emitters from short-living radioiodine as a potentially more serious health threat posed by fallout. Health physicists, radiation ecologists, and others were beginning to understand that though I-131 had a half-life of less than eight days, the isotope moved rapidly through the food chain. After being deposited on pasturage as fallout, the isotope was consumed by cattle who passed it on to humans in fresh milk. Once in the body, some of the radioiodine accumulated in the thyroid gland. Infants and children, it became clear, were especially susceptible to thyroid cancer because of their small thyroid glands and high levels of milk consumption. By 1963 the release of iodine had become, as an internal AEC report put it, "one of the principal, if not the principal, controlling factors in terms of environmental contamination."[24] Nevertheless, as Harold Knapp would comment, at the Nevada Test Site alone, "over 1000 kilotons equivalent of I-131 were released *before* we obtained any reliable data on the I-131 levels in milk in Off-Site communities."[25]

In response to the high levels of I-131 observed in Utah in July 1962, Knapp had begun, on his own initiative, to raise questions about how the radioiodine hazard could be measured. More radically, he sought to "retrofit" new concerns about short-living radionuclides to the incomplete data of the past. Knapp developed techniques for estimating, first, the relation between a single deposit of fallout on pasturage and the levels of I-131 in fresh milk, and, second, the total dose to the thyroid resulting from daily intake of the contaminated milk. According to Knapp's study, the thyroid dose to a one-year-old child from ingested I-131 could be as much as 50 to 250 times the whole body external (gamma) dose to the child if the child consumed at least one liter per day of fresh milk. From a single, relatively low yield test in June 1962, Knapp estimated that the thyroid dose to a one-year-old child in Fruitland, Utah, was likely greater than 130 times that allowed under extant annual radiation protection guidelines.[26]

Perhaps even more disturbing were the questions Knapp raised about radioiodine in what we might call the historical environment of fallout in the 1950s, before significant off-site measurements of it had been recorded. Some of Knapp's earlier work at the DBM had involved radiation "hot spots," areas in which heavy concentrations of fallout are deposited in a highly localized pattern. By extrapolating from external radiation measurements, and relating these converted data to his model for estimating the uptake of I-131 in milk, Knapp showed that, from just one 1953 test, infants who had been living in a "hot-spot" around St. George, Utah (see figure 1.1), could have received I-131 doses anywhere from 150 to 750 times the existing annual permissible doses, substantially elevating the risk for thyroid cancers to develop.[27]

The AEC was hardly eager to publicize either Knapp's findings on I-131 in the environment following the July 1962 Sedan and Small Boy nuclear tests or his retrofitted estimates for the 1953 test. When Knapp sought to publish a report of his findings in the fall of 1962, it was determined to contain "Restricted Data," at least while the AEC scrambled to assemble a special technical review committee and "Ad Hoc Working Group on Radioiodine and the Environment," composed of scientists from Livermore and Los Alamos laboratories, Hanford, the U.S. Public Health Service, and Cornell University, to assess Knapp's paper and its implications for radiation safety practices. Career AEC scientist Gordon Dunning, the deputy director of the Division of Operational Safety—and the author of the DBM study on the biological effects of nuclear war from which Knapp had based his new model for radioiodine uptake[28]—was part of this working group, and for what would amount to perhaps nine months of censorship (as Knapp viewed it), it was Dunning who emerged as Knapp's harshest critic. In Dunning's view, there were, "two basic problems involved, one, the questionable technical validity . . . and two, a policy question of the [Atomic Energy] Commission publishing the paper. In response to a question asked as to the motive for publishing the paper, a member (not the author) of the Commission's staff said, 'The Commission has been telling the world for years that it has been conducting its operations safely, now it appears that this may not be so.' If a member of the staff says this about the paper, what reaction may we expect from the press and the public?"[29] Dunning's main technical criticism of the paper was that Knapp's equations for relating I-131 deposition to its uptake in fresh milk had too many uncertainties—he had extrapolated from too few data points. Dunning also criticized potential errors in data measurements and reliability.[30] Dunning thus called into question, on the one hand, the very mathematical techniques that he himself had developed for the more abstract context of a hypothetical nuclear war, and on the other, the AEC's inadequate data measurement, and the reliability of the same data that had previously been used, before the I-131 controversy, to purport the safety of nuclear testing in Nevada.

"All of science," Knapp responded in a trenchant set of comments for Charles Dunham, director of the DBM, "involves relating one set of measurements to an entirely different set of measurements." Besides noting that he was using the best *available* data, Knapp questioned Dunning's sudden skepticism—for in fact Dunning's I-131 model had worked quite well—and at the same time asked why the AEC had such poor data in the first place:

> Why, when the Division of Operational Safety had a hot spot it didn't expect, didn't it find out what the gamma levels were? All one can say is that some measurements were made and the results were reported and used in a straightforward fashion, and all the rough pieces of data fit together very well and consistently to give exactly the levels of I-131 in milk that Dr. Dunning would have estimated on the basis of his own calculations. It looks mighty silly to suddenly become so doubtful just because when the theory is applied, we suddenly realize we have all been missing the obvious in a haystack for over 10 years, and that the public relations impact of this oversight are painful to ponder.[31]

Up to the July 1962 tests, there had been no systematic effort by the AEC to obtain local fallout levels at the same place and the same time as I-131 measurements in milk were obtained.[32] What questions were being asked about radioiodine at the time? For a U.S. Public Health Service laboratory, Sedan had provided the chance to experiment by comparing physical sampling devices to biological ones—beagle thyroids. The thyroid glands of the caged beagles that had been placed below the dust cloud's projected path were harvested, and the results suggested that the iodine inhalation hazard was not only greater than expected but indeed greater than existing equipment could even measure. Although the analysis of the data yielded few firm conclusions other than the very inadequacy of technical instrumentation in fallout monitoring, "an urgent need" was obvious to project scientists "for additional studies directed toward the determination of radioiodine concentration in air which will lead to the development of truly quantitative sampling methodology."[33] Knapp too was interested in developing better quantitative sampling methodologies. But the experiment is nevertheless telling in terms of its research priorities, unasked questions, and a certain social distancing from local places and communities.[34] Rather than monitoring the downwind human environment more extensively to better establish the sorts of relationships that Knapp outlined, the AEC had contracted out an experiment that worked by irradiating and harvesting beagle thyroids for scientific and technical information, even as the *same* (Sedan) test was significantly irradiating the dairying, gardening, and ranching ecosystems and food chains of downwind Nevada, Utah, and beyond.

By the spring of 1963 it was clear, even to Gordon Dunning, that the boundary of security classification placed over the I-131 study was only a temporary and partial solution to the problems posed by Knapp's study. This was partly because Knapp had already disrupted that boundary, both metaphorically and geographically, by consulting with scientists from the Public Health Service and the University of Utah, along with other people in downwind communities. Dunning noted: "Through meetings and exchange of letters between the author [Knapp] and others outside the Commission, the paper probably is well known. The author spent several days traveling and talking with many people around the Nevada Test Site, followed by exchange of letters with them. This has just goaded them on. We have spent years of hard, patient effort to establish good and calm relations with the public around NTS. Such action as the author's has been harmful."[35] Whether Knapp's or the AEC's public information practices were the more harmful was a matter of perspective. But Dunning was particularly troubled by Knapp's breaching the boundary between the classified and the public sphere and the disruption to the customary movement of information that this breach had created.

"Consider the author's problem," Knapp replied to Dunham in response to Dunning's latest review of his work. *Dunning* had denied him access to relevant AEC fallout data, Knapp argued, and so he had turned to outside public health scientists in Utah instead (including Robert Pendleton), who shared their data with him.[36] In the process, Knapp learned that the Utah Department of Health had "found it necessary to build duplicate monitoring facilities beside those of the Atomic Energy Commission's Off-Site Radiological Safety Organization, because data pertaining to the safety of the citizens of Utah was not forthcoming from the AEC."[37] Knapp also spent a week visiting sites where data had been collected to speak with the data collectors firsthand about sampling procedures, and he sought out additional observations from nearby Nevada ranchers.

While some members of the reviewing committee saw no reason to hold up publication of Knapp's report, Dunning remained concerned that if the AEC published it—even together with a "disclaimer" report by the ad hoc working group that might stress the many uncertainties and qualifications of Knapp's paper, as some colleagues had advocated—then the AEC would be placing itself "in the untenable position of condemning itself," while, at the same time, adding "dignity and prestige to [Knapp's] report." Instead, Dunning proposed "one possible 'out' to the situation":

> Let the Commission tell Dr. Knapp in a matter-of-fact and bland manner that the Commission interposes no objection if he, as an individual scientist, wishes to publish his paper. . . . If Dr. Knapp can find a reputable scientific journal to accept the manuscript, we may expect letters and inquiries. We can treat these in the same professional and unemotional way that we have many others in the past. By doing so, we can make it clear that the Commission has been conducting its operations in a responsible manner and that the Commission has a highly competent staff that has considered Dr. Knapp's ideas.[38]

The AEC eventually agreed to publish Knapp's report (together with the comments of its ad hoc committee review), and Knapp later did find an outlet for a summary of his I-131 study in the journal *Nature*.[39] Before that, however, Knapp found a more public voice for his concerns in August 1963, when Congress's Joint Committee on Atomic Energy reconvened its hearings on fallout, radiation standards, and countermeasures. By this time Knapp had left his position in the DBM's Fallout Studies Branch to become a systems analyst at the Pentagon, where he completed his I-131 study. "Born classified" into an internal AEC controversy, Knapp's report now entered the public record as an appendix in the JCAE hearings, along with both the review team's comments and Knapp's comments on his reviewers.[40] As the public debate over the effects of hot spots and low-level radiation from fallout ran into debates over the ratification of the Limited Test Ban Treaty, Knapp's study—with its grim revisionism of estimating risks for dangers that the public had already been exposed to—joined the discourse of regulation and reform that pushed nuclear testing underground soon after.

Meanwhile, Plowshare scientists and technocrats had been left to wonder: was it really *Sedan's* radioiodine that had caused all these problems?

SMILE, PLOWSHARE

While most of the AEC's public information strategists were concerned with lessening the commission's culpability for the problem of radioiodine contamination, its Division of Peaceful Nuclear Explosives (DPNE) director John Kelly was especially concerned with the bad press that the Utah radioiodine controversy might bring to Plowshare. He described the predicament to Livermore associate director Roger Batzel in August 1962: "The furor about the amounts of iodine in milk has become a real problem and is having some influence on matters related to the Plowshare Program. In this regard, Sedan is being blamed by some officials for a large part of iodine-131 that appeared in milk in Utah shortly after the event. The explanation that although a small part of that fallout might have been due to Sedan the balance resulted from weapons testing during the same period is not well accepted." Kelly conceded, "This is quite likely due to the fact that specific data are not available to substantiate the explanation." The danger, as Kelly saw it, was that Sedan—a 30-percent fission explosive that had itself vented about a full kiloton of radioactivity beyond the immediate vicinity of the crater—might be "unfairly" identified with radiation fallout controversies and, in turn, might jeopardize future Plowshare cratering tests. Kelly's solution was to commission a study, one that might provide "pertinent data and information on the amount of iodine-131 fallout that might be attributable to the Salt Lake City area by Sedan."[41] Although it was not possible to separate the peaceful from the military isotopes in the infants' thyroids, perhaps the boundary could be delineated on the data, and Plowshare could be extricated, to a degree, from the Test Site's downwind geography.

"Smile," wrote University of Utah health physicist Charles Mays to Richard Hamburger, the new assistant director of the AEC's Division of Peaceful Nuclear Explosives in June 1963, "the enclosed calculations support your contention that Sedan was not the major source of I^{131} in the July 1962 Utah incident!" On the basis of three sets of measurements, Mays had determined that Sedan had accounted for only between 10 and 30 percent of the total radioiodine exposure. A fourth set of measurements, which had attributed 50 percent of the radiation exposure to Sedan, had been thrown out of the study because "(a) it is only based on 3 samples and (b) the samples were lettuce, not milk."[42] This excluded fourth data set, however, helps to underscore a key point: whatever the actual total percentage of radioiodine in Utah in July 1962 originating from Sedan, such data did not reflect the geographical variation of radiation levels at the different milk or vegetable collection sites. The average total intake of I-131 during July and August 1962 had been 58,000 picocuries (pc) of radiation (still twice the Federal Radiation Council's *annual* guidelines for young children). But one Salt Lake City dairy measured a total intake (per person at one liter of milk consumed per day) of 117,000 pc; another dairy in the same metropolitan region indicated only 3,400 pc.[43] For Sedan alone, although Pendleton's milk samples ranged from 10 to 2,600 pc of I-131 per liter, Knapp later found that I-131 levels in Utah were also recorded in the range of 10,000 to

100,000 per liter, and noted that "the milk from the locations where the I-131 levels are estimated to have been highest was either not measured at all or not measured soon enough following the shot to give direct experimental verification [to his conclusions]."[44]

Whether or not Mays's news was worth smiling about, his enthusiasm had its limits. For Mays, radioiodine in the environment was a public health problem, not a public relations problem, and Sedan's contribution to it still raised serious concerns. Even if, like many Utahans, Mays seems not to have questioned the practice of nuclear weapons testing as a military and geopolitical necessity, one that justified certain sacrifices, it was difficult for him to see why *Plowshare* tests were so necessary. As Mays put it to Hamburger, "the issue is not whether Sedan was the main culprit: it is whether a population exposed in excess of the radiation protection guide by fallout from essential weapons testing, should be further exposed by Plowshare devices which are *not* essential to National Survival." This was an argument that neither Hamburger nor Kelly nor the Plowshare scientists at Livermore wanted to hear. "Your interests in reducing the radiation hazard from Plowshare devices," he concluded his letter, "and mine, in reducing the radiation exposure to Utah children are in many respects compatible."[45] Mays's letter also made clear, however, that in other respects the interests of Plowshare and of Utah public health advocates were directly opposed.

PUBLIC HEARINGS

Partly in response to the radioiodine controversy, the JCAE convened fallout hearings in June and August 1963. Even though the scientific controversy was by no means settled, here at least the terrain of debate—set forth not only by those like Gordon Dunning but also by the former insider Harold Knapp, the Utah health scientist Charles Mays, Caltech geneticist E. B. Lewis, the Committee for Nuclear Information's Eric Reiss, and many others—could be laid out in public view.[46]

Mays began his testimony by speaking about radioiodine in Utah from the July 1962 tests and then shifted to the question of radiation exposure from Nevada tests during the 1950s. Mays estimated that thyroid doses of I-131 for Utah infants might have been as high (on average) of 6.3 rads in 1953 and 8.6 rads in 1957. In all, he estimated, with various qualifications, that "approximately 250,000 Utah children have been exposed to crudely estimated average thyroid doses of 4.4 rad prior to age 2," or more than four times extant dose limits. Yet the magnitude of this overexposure did not compel Mays to directly challenge the AEC's nuclear testing practices. "Continued weapons development," he argued, "is essential for the safety of this nation. Some radiation exposure is a small price to pay for liberty. It would indeed be poor economy to lose a nuclear war because our weapons testing program had been needlessly hampered. I believe that no test needed for national defense should be canceled." Without viable claims that geographical engineering was similarly necessary for "national survival," Plowshare advocates found little to smile about in Mays's statements before Congress. For Mays, echoing the pragmatism and blunted criticism that marked the opinions of many at the hearings, the question of how to reduce I-131 exposure in the future, given the logic of deterrence and purported necessity of continued testing, was "by far the most important [issue] because there is little we can do about previous incidents except to learn from them."[47] President Kennedy himself sought to reassure the public following Mays's testimony, noting that there was "some controversy" about the reports and that "we do not believe that the health of the children involved has been adversely affected." But as Mays responded to the president's comment, the reason no evidence had "come to light" of damage to the children's health was that "the proper study has not been made."[48]

Learning from past incidents and making recommendations for testing more safely in the future were also primary goals of the ad hoc working group on radioiodine and the environment that had been convened, under Gordon Dunning's leadership, across the AEC's classified spaces following Knapp's initial report on the radioiodine hazard. The group, as Dunning described it, "addressed itself only to the technical aspects of the problem, i.e. what is needed to get the job done correctly."[49] Their recommendations—prodded, after Sedan and the Utah radioiodine controversy, by new knowledge of an I-131 food chain that was much closer to home for most Americans than Sr-90 in caribou—included the extension of radiation sampling to a distance of five hundred miles from ground zero, increased cloud sampling, and more extensive monitoring of radiation in milk in downwind areas. The need for greater cooperation with public health agencies was stressed, along with the need for more scientific studies of the radioiodine problem. No new fallout standards were set. But in this fallout-sensitive political climate that Sedan had helped to produce, ongoing debates over the ratification of the Limited Test Ban Treaty nevertheless threatened to restrict future progress in nuclear earthmoving unless Plowshare advocates could succeed in undermining progress in the international regulation of nuclear testing first.

New Limits: Plowshare and the Limited Test Ban Treaty

If Plowshare had been an irritant to discussions of a nuclear test ban treaty in 1958, then in 1962 and 1963, when a new treaty that would limit nuclear testing to underground explosions was gaining currency, the category of peaceful nuclear explosions (PNEs) became a potentially more divisive issue. International negotiations of a comprehensive test ban had continued between 1958 and 1962—for some 353 meetings—under the Special Geneva Conference on the Discontinuance of Nuclear Weapons Tests, bogged down by issues of compliance and verification, site inspections (which the Soviets categorically rejected), and more generally, mutual mistrust. President Kennedy maintained a commitment to continuing negotiations amid the turbulent geopolitics of the times. Under Kennedy's leadership, the idea emerged among American arms controllers for a *partial* or *limited* treaty, more palatable to hardliners in both the United States and the Soviet Union, that would respond to scientists' and activist groups' concerns over the health effects of fallout by permitting only underground tests, forbidding them in the atmosphere, underwater, or in outer space. Although Kennedy still sought a comprehensive test ban treaty, he hoped that a Limited Test Ban Treaty (LTBT), as it came to be called, might provide a practicable first step in controlling testing (and perhaps the arms race more broadly) and a signal of lessening Cold War nuclear tensions.

Where would such a treaty leave Plowshare's nuclear excavation program? In March 1962, a joint UK-U.S. LTBT was drafted by the U.S. Arms Control and Disarmament Agency for presentation at a new eighteen-nation disarmament committee convening in Geneva. The treaty, which was submitted together with another that banned nuclear tests altogether, included a provision—a nod to Plowshare, with an eye toward future international contracting opportunities—allowing all states to realize the future benefits of PNEs, so long as there was unanimous consent of the United States, United Kingdom, and USSR. Soviet negotiators in Geneva, however, had little enthusiasm for any PNE exception, which they regarded as a scheme for acquiring military information from "peaceful" explosions. Both treaties were rejected, followed by a round of nuclear saber rattling that included a colossal, 50-megaton Soviet atmospheric test and more than 20 megatons (combined) from some forty U.S. atmospheric tests in the Pacific.

After the resolution of the October 1962 Cuban missile crisis, however, both Kennedy and Premier Nikita Khrushchev saw a window of opportunity, and perhaps of necessity, to accomplish something in the area of arms control and nuclear testing regulation. As formal talks resumed in Geneva in 1963, the AEC once more pressed for provisions to allow for continued testing in nuclear excavation. U.S. negotiators suggested unprecedented international inspection and verification procedures as one means of allowing excavation tests, but soon found the boundary dividing the peaceful explosives from the nonpeaceful ones to be rather slippery. As AEC chairman Glenn Seaborg explained it to Under Secretary of State W. Averell Harriman, for nuclear excavation to achieve a reasonable "degree of cleanliness . . . we must use the most advanced device design principles known to us. The principles are identical to those used in our most advanced weapons, thus, these devices cannot be disclosed to other nations." In place of actual device inspection, Seaborg proposed more limited forms of observation "that would not present insurmountable obstacles" to Plowshare, including access to fallout debris.[50]

For others, however, the real problem was that Plowshare was becoming an obstacle to the treaty process, not the other way around.[51] When Kennedy and Khrushchev elected to sidestep the eighteen-nation Geneva conference and to proceed instead with direct, three-party talks in Moscow, Kennedy included no one from the AEC nor any of its supporters in the JCAE in the nine-person U.S. delegation, which did include two science advisors. Harriman, a former wartime ambassador to Moscow whose diplomacy had helped to make the meetings possible, was appointed by Kennedy to lead the U.S.-UK delegation. He was sent to Moscow in July 1963 with instructions to "get an agreement and come home"—preferably a comprehensive test ban treaty but at least (and more likely) a limited one.[52]

As Harriman relayed one conversation in his nightly report to the White House, Khrushchev himself expressed keen interest in nuclear earthmoving techniques for canal building in Central Asia and the Arctic, but feared that the inclusion of a PNE provision would look like a loophole to the international community. An agreement on PNEs, Khrushchev conveyed to Harriman, was better reached later, through a treaty amendment, when nuclear tensions had been reduced by a LTBT. Informed that the Soviets would accept the treaty if the PNE article was dropped, Kennedy was compelled to assess the value, and purported future benefits, of Plowshare's nuclear earthmoving program in relation to the broader goals of the treaty. For advice, Kennedy did not turn to Seaborg, who was informed of the treaty only after it had been signed, but more likely to the words of his science and technology advisor, the MIT (and former Manhattan Project) engineer Jerome Weisner, who had informed Harriman before his departure for Moscow that nuclear excavation capabilities were "years in the future at best."[53] Kennedy agreed to drop the PNE provision, with the tacit understanding that the treaty could later be amended, but as part of the bargain insisted on a clause permitting withdrawal from the treaty. A withdrawal clause, the administration believed, would increase chances of ratification of the LTBT in the Senate. After six years of negotiations in Geneva and twelve days in Moscow, Harriman returned to Washington with a treaty and what was, testing opponents and arms controllers hoped, a step toward a comprehensive test ban treaty in the future.

The final text of the LTBT, signed by the United Kingdom, United States, and the Soviet Union on August 5, 1963, made no explicit provisions for PNEs.

Plowshare tests and projects were thus placed under the same guidelines as weapons tests. Signatory states agreed not to carry out "any nuclear weapon test explosion, or any other nuclear explosion at any place under its jurisdiction or control," either

> A. In the atmosphere, beyond its limits, including outer space, or underwater, including territorial waters or high seas; or
> B. In any other environment if such explosion causes radioactive debris to be present outside the territorial limits of the state under whose jurisdiction or control such explosion is conducted.[54]

The agreement presented a major setback for Plowshare's supporters. First, since nuclear cratering explosions only began underground, they might be interpreted as banned "atmospheric" explosions. But more ominously, the restriction on radioactive debris moving beyond international boundaries represented, for Plowshare, a sweeping new legal and geographical limit.[55]

Alongside the final phase of the JCAE's public fallout hearings that same summer, the Senate Foreign Relations Committee held both secret and public hearings in August over the ratification of the LTBT. In his testimony over the LTBT ratification, Edward Teller gave little weight to revelations from the radioiodine controversy, or to the seriousness of the fallout hazard in general. "These two questions, the test ban and fallout, are linked only by propaganda," he maintained.[56] Teller opposed the treaty, as he reiterated for a national television audience shortly after the hearings, because it limited knowledge. "I don't want bigger explosives," Teller explained, "I do want knowledge, knowledge that comes from testing, knowledge to be applied for our defense, knowledge to be applied for the peaceful use of nuclear explosives. In the way of increasing this badly needed knowledge, I think the more we have the better, and we can do it cleanly and without disturbing anybody in any serious sense. As far as knowledge is concerned, more and more will be needed." What is more, Teller suggested, without a vigorous Plowshare program, what would prevent the poorer countries from turning to Soviet PNEs for their future development projects?[57]

And yet while some Livermore spokespersons, including Teller, opposed LTBT ratification, the majority of Los Alamos scientists favored ratification, according to its director, Norris Bradbury.[58] For Seaborg and the AEC, already under increased scrutiny due to fallout controversies, there was little choice but to join Kennedy in supporting ratification. It was clear to Teller that if the LTBT was to be ratified, nuclear warhead testing would not survive under any circumstances, but he hoped to loosen the constraints that had been placed on the Plowshare program. Teller called upon the senators to enact a treaty article that made exceptions to allow for continued progress in PNE testing. Without such a change, Teller insisted, the LTBT would "make the difference between a most effective Plowshare and practically no Plowshare at all."[59]

When the LTBT was ratified, Teller's pessimistic reading of the treaty and the future of geographical engineering was shared by one Plowshare scientist at Livermore, who scrawled "All Harbors and Canals involving Oceans are out," "Australian Harbor Out," and "Panama Canal out unless amendment" in the margins of his advance draft of the LTBT.[60] But with the treaty more or less settled and a new regulatory environment taking shape, it was time for Plowshare's scientist-planners to shift their attention from the politics of treaty negotiations to the subpolitics of definition. That is, instead of asking how the treaty could be reshaped to accommodate nuclear excavation, the AEC would now ask how the excavation program could be recast but still vigorously pursued within the terms of the treaty.

In this sense, the clause that prohibited the presence of radioactive debris beyond international boundaries was now, for the AEC, "open to some interpretation."[61] Rather than an absolute limit, Seaborg adopted an interpretation of the "presence" of radioactive debris that was based on a quantitative standard for small concentrations of radionuclides. Only concentrations that would have "some practical significance, such as an effect on human health," were to be considered "present" under the treaty terms.[62] Although Seaborg had not managed to safeguard nuclear excavation during treaty talks, under this rather optimistic interpretation of the treaty he provided Livermore scientists with a philosophical position from which to proceed: larger projects such as the sea-level Central American canal would have to await an LTBT amendment, but most of the cratering experiments and demonstration projects on its agenda in 1962 could still be carried out. Teller's autopsy of the program had been premature.

Five-Year Plan

At the same time that Plowshare's planners were encountering ever more constricting obstacles, they found new allies and increased funding for the excavation program, principally in connection with the quietly developing Central American canal project. The remainder of this chapter examines how Plowshare scientists understood and negotiated this paradox, incorporating both new limits and new resources into experimental planning as well as public information strategies, leading ultimately to the completion of additional nuclear cratering tests. However much Plowshare scientists felt stifled by "political" issues, they never seem to have doubted that their experiments would be turned into a useful applied science. Their optimism depended less on experimental success than on sheer resilience.

At the dawn of the program in 1957, Livermore scientists had hoped to enact a "useful demonstration project" within eighteen months. By the spring of 1962, however, still without a serviceable demonstration in nuclear earthmoving, it was clear that plans for the experimental program required significant revision. At that time, in connection with the embryonic Central American canal study, the AEC established "the goal to achieve a nuclear excavation capability within the next five years." The promise of an economically produced sea-level Isthmian canal remained, for Plowshare supporters in Congress, the main foreseeable benefit to the nuclear excavation program. In developing their experimental program together with AEC division managers Kelly and Hamburger, Livermore's Plowshare scientists thus crafted a five-year plan (mirroring Stalinist five-year plans in its ambition) that would, if accomplished, allow them to master the art of nuclear earthmoving. They outlined a tentative experimental program aimed at reducing both the biological and the economic costs of nuclear excavation, arguing that their cratering tests would provide "a scientific and technical basis for an information program to establish public and official acceptance, both domestically and internationally, of nuclear excavation techniques." The new experimental program included six different kinds of nuclear excavation tests, represented here in the summary of the five-year plan in table 4.1. From a single cratering shot ten times larger than Sedan to blasts in different rock types to more complex combinations of simultaneous and sequential detonations to, by 1967, at least one "clearly useful" demonstration project, the five-year plan was both systematic and ambitious.[63]

Initially two Livermore projects were to be pursued: Schooner, a cratering experiment much like Sedan but to be conducted in hard rock; and Canoe (later renamed Buggy), a simultaneous row-charge of five nuclear blasts that would provide new data on linear channel and canal excavation, data that Sedan, as a replacement for Chariot, had not produced. From there, Plowshare would move on to earthmoving schemes of greater scale and complexity. Some of these experiments and demonstration projects were to be carried out away from the Test Site, such as Project Galley, an experiment exploring the combination of effects engendered by a row-charge explosion "done at a site near water," for which Livermore scientists were already eyeing Australia's long coastline. But while Plowshare scientists were unflaggingly optimistic about future progress in nuclear excavation, given adequate opportunities to experiment, program planners saw potential obstacles ahead involving issues of "public relations." Regarding public safety, for instance, Plowshare planners insisted that the chief problem was actually that of safety *criteria*. "The lack of such guides played a part in the inability to categorically establish the absence of any hazard in the Chariot project."[64] Without clear guidelines in place defining permissible

Table 4.1
An Excavation Capability within Five Years? The Experimental Program in Nuclear Excavation, July 1962

Type of experiment	*Description of purpose*	*Projects planned*
Scaling	Designed to "extend knowledge of scaling and entrapment of radioactivity from the kiloton range to the hundreds of kiloton and megaton ranges." Experiment could possibly be conducted as "a useful or quasi-useful project."	Project Phaeton: megaton yield explosion planned for fiscal year (FY) 1965.
Media effects	To "develop information on scaling and entrapment of radioactivity in at least two types of hard rock."	Project Schooner: 50–100-kt explosion, planned for FY 1964 or 1965, to be detonated in basalt. Site selection studies under way. Project Dogsled: 10 kt explosion to be detonated in either sandstone or limestone, possibly at Utah's Dugway Proving Grounds.
Row charge effects	To extend knowledge of "row charges," i.e., the simultaneous detonation of several explosives placed in a row so as to excavate a channel or canal.	Project Canoe (later renamed Project Buggy): simultaneous detonation of five 5–10 kt explosives at the Nevada Test Site. Tentatively scheduled for late FY 1963 or 1964.
Combination of effects	To combine "scaling, media, and row charge effects in uneven terrain seems to be indicated." "Preferably this should be done at a site near water so that flooding effects could be studied. This might be somewhat similar to Chariot."	Later code-named Project Galley: to be conducted in FY 1966.
Multiple operations	To extend knowledge and experience with more complex cuts, in preparation for the sea-level Trans-Isthmian canal, which would require arrays of lateral and vertical charges involving both simultaneous and sequential detonations.	Army Corps of Engineers to investigate these phenomena through an experimental high-explosives program. Nuclear experiment would be pursed for FY 1966 or 1967.
Demonstration	Public demonstration of at least one "clearly useful project."	To be completed by 1967, prior to canal construction.

SOURCE: AEC "Plowshare Excavation Program," July 10, 1962, AEC 811/105. DOE doc. NV0075573, Nuclear Testing Archive, Las Vegas, Nevada (NTA).

levels of radiation exposure from nuclear cratering blasts, the nuclear excavation program would continue to encounter public skepticism.

In addition to nuclear tests like Schooner, Buggy, and Galley, the five-year plan called for a public relations strategy that would work on two levels: first, the aim was "public acceptance" through positive reports of experimental results and through efforts to establish more liberal definitions of what counted as safe nuclear earthmoving. On another level, the goal of achieving an excavation capability within five years would also require "user capability," a category that included the production of new technical manuals on nuclear excavation and the cultivation of new partnerships in the work of earthmoving. The Army Corps of Engineers, which had built the first Panama Canal under the authority of the Panama Canal Commission and which remained, in stream management work carried out by its district offices across the United States, the country's most prolific earthmover, constituted one such critical alliance for Plowshare. Anticipating that the corps would be the obvious choice for managing any future nuclear canal project in Central America, the AEC engaged the army engineers in the nuclear excavation program through a parallel program in high-explosives experimentation, eventually leading to the establishment of the corps' Nuclear Cratering Group within the Livermore laboratory, as we shall see in chapter 5.[65] Efforts to establish new institutional relationships also included contacts with academic physics and engineering departments as well as between Livermore scientists and government agencies, from the California State Highway Commission to the Australian Atomic Energy Commission. All of these relations would assist in turning the idea of nuclear excavation into a more or less normal engineering application.

But while the goals of public acceptance and user capability may have been universally accepted among Plowshare supporters, differences emerged over how these goals for an effective public information policy should be achieved. By March 1963 a "vigorous program of public education and information" no longer seemed the best course of action for Plowshare, according to Livermore director John Foster. As the AEC's Dan Wilkes put it, verifying a previous conversation with Foster, "My impression is that AEC works harder at promoting Plowshare than you figure they should. [Rodney] Southwick and other AEC information people may be overenthusiastic and may stimulate publicity, through misunderstanding, that may be undesirable."[66] After the ratification of the LTBT, with the question of exactly how nuclear excavation would fit into the new testing regime still very much in flux, Foster's advice regarding a less vigorous public information campaign must have made a good deal of sense to AEC program planners. It must have been equally clear to Plowshare scientists at Livermore that carrying out the experimental plan—calling for five excavation experiments by the end of fiscal year 1965, and at least three more by 1967—would not be easy.

Faced with the practical task of setting the experimental plan for nuclear excavation in motion—at precisely the moment when Utah public health scientists like Charles Mays were considering the necessity of Plowshare to "national survival"—Livermore and AEC leaders, in the spring of 1963, began to raise questions about how the plan should be revised. "The problem," as DPNE director Kelly saw it, was how to "resolve the conflict between the reduced rate of clean weapons development and the Plowshare excavation program requirement for clean devices."[67] Although plans had initially called for carrying out both Project Buggy and Project Schooner in late 1963 or in 1964, Livermore's Foster now conceded that sufficiently "clean" explosives would not be ready in time.[68] The alternatives for the excavation program, as Kelly explained to the AEC's general manager, were thus either to support additional underground "clean device development"—beyond that already funded by the Division of Military Applications—or to "accept the current schedule of clean device availability" and go ahead with Buggy and Schooner as planned with the "cleanest available" explosives. This latter option, Kelly observed, had "the multiple disadvantages of releasing larger amounts of radioactivity, possibly creating negative public reaction to nuclear excavation, and certainly not demonstrating low environmental contamination."[69]

The solution, proposed by Kelly and supported by the DMA, split the difference between these alternatives. First, seven new device-development tests were requested and authorized.[70] Although much of the data produced by these experiments would also provide a resource for military "clean explosives" research, the device-development tests were to be "justified and announced as Plowshare experiments independent of the weapons program."[71] Second, Buggy would be postponed for at least one year. Third, Project Schooner, designed to blast a Sedan-sized crater in hard rock, was to proceed on schedule using the "best available."

Mounds and Craters

MAIN STREET, IDAHO

In March 1963, eight months after Sedan, Livermore and AEC scientists scanned the landscapes of northern Nevada and southern Idaho as they toured the area by helicopter, geological maps presumably in hand, in search of an appropriate site for the Plowshare program's next peaceful nuclear explosion. Project Schooner was to be another 100-kiloton nuclear-cratering blast, but this time conducted in either granite or basalt to provide new data on the mechanics (and hazards) of cratering in hard rock.

It was the search for an ideal hard-rock surface—rather than a renewed effort to enact a useful public demonstration project (since Schooner was designed, like Sedan, to produce only another hole in the ground rather than a useful engineering structure or a model of one)—that had pointed the scientists in their chopper to the region. Located just several hundred miles north of the Nevada Test Site, the area was already home to the Saylor Creek Aerial Gunnery Range and the nearby Mountain Home Precision Bombing Range as well as the AEC's National Reactor Testing Station in southeastern Idaho. In their site selection work, scientists sought to fulfill four main criteria:

1. A large mass of dry, relatively uniform hard rock extending at least four miles horizontally, and at least one thousand feet deep;
2. A relatively flat area, with line-of-sight viewing possible at five miles from ground zero;
3. Large accessible sector for fallout in direction of reasonable wind expectancy;
4. Remoteness from population center: no towns or villages occur within twenty miles.[72]

The reconnaissance mission collected a wide range of spatial data, ranging from geology to land uses to censuses of people and dairy cows in the region. On the basis of some calculus involving these data, along with the interrelations of national and regional politics that had already produced a militarized and nuclear landscape in the region, the first site selected for Schooner was near Saylor Creek in southwestern Idaho. A preliminary fallout sector was mapped to a (rather conservative) distance of 150 miles downwind (see figure 4.3). The quadrant pointed south from ground zero and included several ranches within twenty-five miles and three towns at roughly fifty miles' distance on the Nevada side of the state border: Jarbridge, Jackpot, and Owyhee, the last the largest village (population 150) on a Western Shoshone Indian reservation. Along with scores of sheep and cattle ranches located inside the fallout sector, another 800 people lived on the reservation outside of Owyhee, bordering on the fallout sector.[73] Although a subsequent AEC review panel found this fallout sector unacceptable, the proposed Schooner site was shifted just twenty miles southwest to a location between the east and west forks of the Bruneau River—maintaining a southward fallout sector fairly similar to the one illustrated in figure 4.3. Here, located on a deep granite formation apparently unmatched at the Nevada Test Site, preliminary geological and engineering field studies were initiated by AEC contractors.[74]

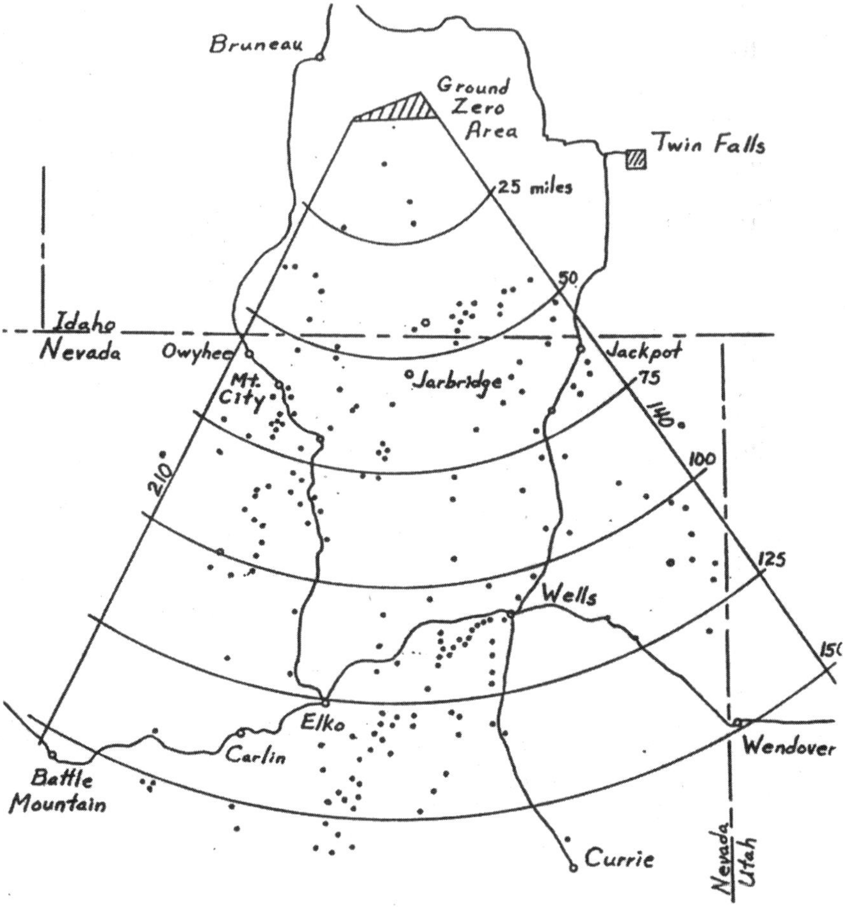

Figure 4.3. Preliminary proposed fallout sector for Project Schooner at Saylor Creek, Idaho. *Source: Public Health Service, "Preliminary Site Surveys—Project Schooner," June 11, 1963. Plowshare Box P-5, Lawrence Livermore National Laboratory.*

In their early plans for Schooner, Livermore and AEC scientists and public officials pursued a strategy designed to make Schooner appear publicly accessible. Schooner was designated an "open" event, a technological spectacle with official observer and news media programs that would include a visitor's bureau and an "informational annex" near the site where unclassified public information would be made available. At the same time, the AEC hoped to restrict access to the experiment itself. Ecological studies in particular were to be more far more circumscribed than was the case for Chariot in Alaska: modest pre- and postshot inventories of radioisotopes were to be the primary focus, extending to a downwind distance of just forty miles, and merely ten miles in the freshwater ecosystem.[75]

As important as these physical parameters was the question of who would carry out the ecological surveys. For this task the AEC turned to familiar Nevada testing contractors in the Public Health Service and the engineering firm ReeCo rather than to teams of local university scientists and regional experts like those who had contributed to the Chariot studies. All contract scientists would work through AEC Office of Information channels. Through such careful specifications, the AEC sought to fashion a rigorously controlled experimental space out of the Bruneau River site, a space where Livermore's Plowshare scientists and Nevada testing personnel would maintain unmistakable authority over the evaluation of the experiment.[76]

A public information plan for the experiment was designed to minimize "unfounded concern that may arise from public misunderstanding of the Project Schooner operations."[77] Judging from early encounters with local officials, the AEC anticipated few such concerns from nearby communities. After an official AEC announcement that the Saylor Creek area was "under investigation" for possible use for a Plowshare experiment, spokespersons from the AEC Office of Information visited with local officials, chambers of commerce, ranchers, and newspaper editors. As the AEC public relations team saw it, Schooner received a warm response:

> Generally, residents of the area appear to be enthusiastic at the possibilities of using the Saylor Creek site for this Plowshare event. The area as a whole is an economically depressed area. . . . Project Schooner is viewed as a possible economic boon to the state of Idaho and especially to the area covered in this report. Residents also view the project as a vehicle for obtaining favorable national attention and publicity. At both Gooding and Mountain Home, city officials indicated that they would favor detonating the experiment under Main Street if it would prove to be an acceptable site.[78]

AEC officials, who were more than happy to promote the nuclear earthmoving experiment as economic development, were reassured to observe "little or no concern over possible radiological or seismic-motion effects of the proposed project." Few questions were raised about the "possible release of radiation." It was noted that one rancher "in the Shoshone area" wanted to be assured that radiation would not affect his crops and livestock, but also that "another rancher, in the Hagerman area, questioned the validity of many published reports relating to the harmful effects of radiation."[79]

It is difficult to predict what might have become of this localized controversy between the Hagerman and Shoshone ranchers in the event that the

AEC had proceeded on schedule with plans sketched out as late as May 1963, when Schooner remained atop Plowshare's experimental agenda even as Livermore and AEC planners were gently shifting the direction of the excavation program toward underground device development. Whether local agriculturalists, interested environmental scientists, Native American activists, or Interior Department bureaucrats, opponents of Schooner might have emerged to demand more extensive ecological studies to precede the test; and in doing so, they might have drawn further attention to the amount of radioactivity already moving through the ecosystems of the region as a result of more than a decade of nuclear testing at the Nevada Test Site to the south. Or perhaps history would have repeated itself more brutally for the Shoshone and Paiute people, living on both formal and informal reservations across Nevada's basin and range landscape, whose lands had been already condemned for government purposes and whose environments (on which many depended for the wild game that was a crucial part of their resource base) had been heavily contaminated.[80] In the event, though, it was in Washington, D.C., not Owyhee, Mountain Home, or Livermore, that the future of Schooner was to be decided.

WASHINGTON, PANAMA, AREA 18

Determined not to repeat the mistakes of Chariot, the AEC wanted publicity for Schooner and the Saylor Creek site kept to a minimum during the experiment's early planning phases. In late August 1963 all personnel associated with the project were instructed to refrain from making any public comments indicating that Saylor Creek had been selected for Schooner, and "official visits" to the project site were reduced.[81] These cautious politics on the part of the AEC were hardly surprising given ongoing congressional hearings over fallout hazards and the ratification of the LTBT. Nor could it have been a great shock to Plowshare scientists, given these national and international political contexts, when, in early October, AEC leadership in Washington decided to postpone Schooner for the time being.[82] The AEC had been made aware that it stood no chance of gaining President Kennedy's approval to detonate any large-scale excavation shots so soon after the LTBT's ratification.[83]

For Plowshare's resilient scientist-planners, the new situation meant that, given the constraints imposed on Plowshare by the LTBT and the partial redirection of emphasis toward device-development tests, small-scale cratering experiments rather than large ones like Sedan and Schooner might be used to keep the excavation program moving forward, at least to an extent. Instead of the 100-kiloton nuclear-cratering test in Idaho, Livermore proposed a rather conservative 100-*ton* "pilot experiment" for Schooner, code-named Project Sulky, for early 1964.[84] "Concurrent with the development of improved explosives,"

Livermore Plowshare scientists now argued, "it is essential to proceed actively to increase the understanding of excavation technology." Sulky was the smallest explosive yield that could still produce useful experimental data. While Schooner, they claimed, was still necessary at an "early date," Sulky was to be carried out in advance in the basalt caprock of Area 18 at the Nevada Test Site, designed to aid in Schooner's experimental design.[85]

Even little Sulky would face delays as the testing program responded to broader concerns and controversies over the implementation of underground testing regimes and the procedures and technologies designated for their verification. After the Utah (and Washington) radioiodine controversies of 1962–63, the ratification of the LTBT, and the move from Schooner to Sulky—which replicated, in some ways, the geographical retreat from Chariot to Sedan before it—it was not a promising moment for the excavation program.

But Livermore was patient with Plowshare, and new knowledge of radioactivity hazards had done little to shake the laboratory's confidence in geographical engineering. In December 1963, Harold Brown, the former Livermore physicist who was by this time director of defense research and engineering at the Pentagon, asked Gerald Johnson to comment on health and safety aspects of Livermore's February 1960 nuclear canal scheme, which had called for 325 nuclear explosives, totaling 310 megatons of explosive energy. Johnson reflected that "the health and safety problem from fallout, in the light of experience up to that time [February 1960], seemed manageable using proper operational procedure with the cleanest explosives then available. At that time, this fallout was considered by those making the study to be acceptable from a public health standpoint." How had the questions raised over Chariot, and those raised about radioiodine by Harold Knapp and others, changed this standpoint? Modestly. Johnson continued: "Now, however, in areas where fallout can occur on dairy pasturage, some levels would not be permitted unless the milk was diverted from immediate human consumption for an appropriate time. Nevertheless, even this would not affect the feasibility of the canal construction."[86] The entire excavation program, Johnson knew, depended on the credibility of the canal project—now conceived, for the favored Darien Gap route, in a configuration of 310 "devices" with a total yield of 258 megatons—and he and his Livermore colleagues were not going to give up on geographical engineering at this point.[87]

In September 1964, as I noted in the Introduction, Congress authorized a 17.5-million-dollar study to explore the feasibility of blasting a sea-level canal through the Central American isthmus "whether by conventional or nuclear excavation." Unspecified "technical experts," the *New York Times* reported in December, when President Johnson announced in a surprise White House statement that the United States intended both to proceed with plans to build a new sea-level canal and to negotiate a new treaty on the existing canal, "said that the nuclear method would be preferable from the standpoint of economy."[88] The canal bill authorized the creation of a new Atlantic-Pacific Interoceanic Canal Study Commission under the auspices of the Army Corps of Engineers, which would initiate surveys of several potential canal routes and work closely with the Plowshare program. Indeed, the canal commission promptly requested that the AEC complete six to eight nuclear excavation experiments by June 1968—a schedule that matched well with the AEC's own five-year plan—so that it could evaluate the nuclear option in its final report to Congress, due in December of that year. Plowshare's budget was increased to twelve million (1964) dollars, the beginning of a relatively steady climb in support bolstered by the canal study (figure 4.4). This line of funding is perhaps what President Johnson meant when, in December 1964, he pronounced nuclear excavation to be "national policy" in a reassuring missive to Glenn Seaborg.[89] What had Plowshare done to warrant this move from politics to "policy"?

For those in a position to contemplate such ambitions, a sea-level canal offered a great deal to the United States. To pass through the existing fifty-mile canal, a ship (entering from the Caribbean) had to be raised eighty-five feet at the Gatun locks and lowered thirty feet to sea level at the Pedro Miguel locks at the Pacific end of the canal. Even without delays, the trip across the canal took eight hours, but increasingly congestion had forced ships to wait in queue, dropping anchor overnight before passing through. This "canal bottleneck," the *Wall Street Journal* had complained in 1961, might "throttle the expanding international trade of the United States and a score of other nations."[90] In addition, there were strategic, geopolitical concerns with the canal as it stood: modern aircraft carriers were too big to pass through the canal's locks and channels, and the route was vulnerable, according to theorists of nuclear geopolitics, to a strategically costly nuclear missile attack.[91]

As others have argued, the timing of President Johnson's support for nuclear excavation and the canal bill had more to do with a crisis in U.S.-Panama relations than with technical progress at Livermore, whatever the geopolitical and economic urgencies.[92] Early in 1964, following anti-American riots in Panama City and increasing hostilities across the Canal Zone, the Panamanian government demanded a renegotiation of the canal treaty. Johnson agreed to renegotiate the treaty just three months after the passage of the canal bill. But he did so from an improved bargaining position: the sea-level canal, and the infusion of capital and commerce that could come with it, would perhaps be carved in northern Colombia instead of Panama. U.S. negotiators would soon promote a draft canal treaty with Panama that included a curious blend of carrot

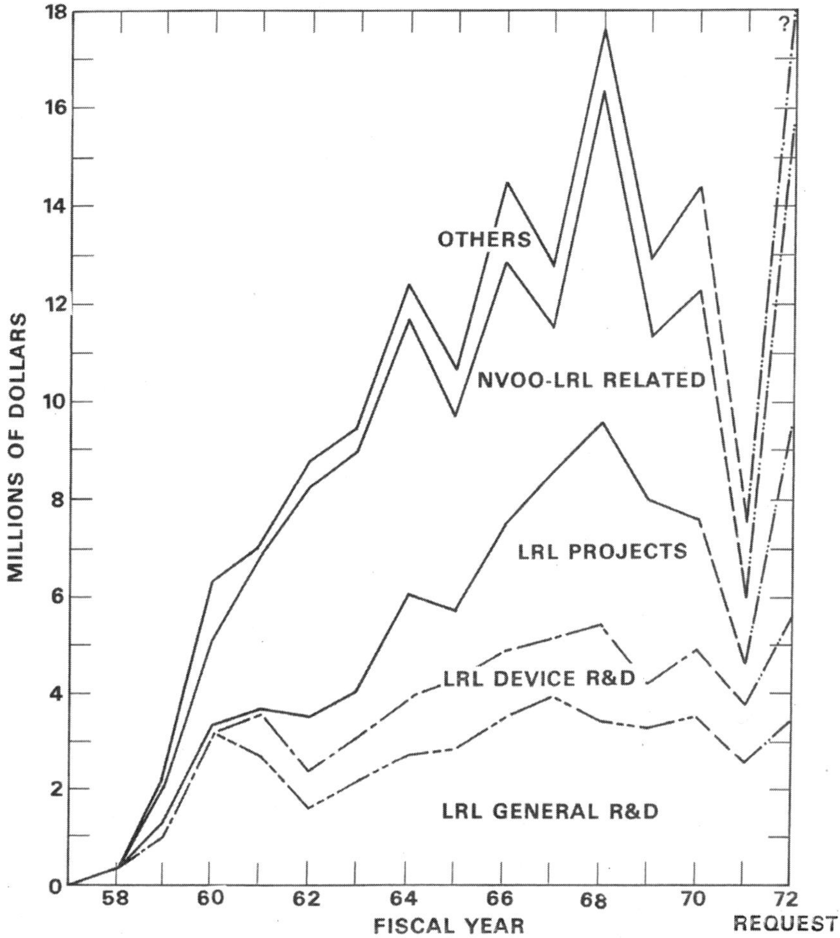

Figure 4.4. Plowshare budget for research and development at the AEC's Livermore laboratory (LRL) and Nevada Operations Office (NVOO). Fiscal year 1971 spending was reduced to $8 million, as indicated by the broken line, but the optimistic projected increase for fiscal year 1972 was not approved by Congress. Plowshare was appropriated approximately $5 million, $7 million, and $6.8 million for fiscal years 1972–1974, respectively, after which the program was virtually phased out. No significant funding was allotted to the excavation program after 1970. *Credit must be given to the University of California, Lawrence Livermore National Laboratory, and the Department of Energy (Atomic Energy Commission) under whose auspices the work was performed when this information or a reproduction of it is used.*

and stick: Panama was offered veto power over the use of nuclear explosives if the new canal was excavated from one of the Panamanian routes under consideration, but not if it was blasted just across the border in Colombia.

For Plowshare's program builders within Livermore and the AEC, the survival—and planned expansion—of the nuclear excavation program at this time was a considerable achievement, given the obstacles that it had faced. Scientists had succeeded in constituting nuclear excavation as a still credible idea, at least for those who wished to believe in it. As President Johnson explained his government's interest in carving a canal across Central America for the second time in the twentieth century (and the possibility of doing so by means of nuclear explosions) in remarks from the White House before television cameras in December 1964: "These changes are necessary not because of failure but because of success; not because of backwardness but because of progress. . . . This new age requires new arrangements."[93]

The same day that Johnson made his White House remarks, experimentation in nuclear excavation resumed. Project Sulky was finally detonated, after numerous delays, on December 18, 1964. Although the original public information plan had called for a public announcement of the test at least twenty-four hours in advance, the AEC later shifted strategies and decided not to make any announcements until twenty-four hours *after* the test; it was argued that, though the radiation release would be quite low from such a low-yield blast, an advance announcement might allow another country to "deploy special gear and thus to detect an otherwise innocuous explosion." Sulky was expected not only to produce new information about the mechanics of nuclear cratering in hard rock, but also, with the explosive buried deeper than the "optimum depth of burial" that would have maximized crater size, it was an experiment in the reduction of radioactive venting through deeper emplacement of the explosive.[94]

As it turned out, venting *was* reduced, but this was because, instead of a crater, Sulky unexpectedly produced a mound of broken rock with a slight depression in the center (see figure 4.5). Had the AEC gone ahead with a planned press release indicating anticipated crater dimensions of roughly thirty-two feet in depth with a radius of sixty feet, Sulky might have proved quite embarrassing.[95] Instead its public relations goals were facilitated by the rigorous policing of information flows: the initial press release two days after the explosion briefly described the mound of broken rock at Area 18, but made no mention of prior cratering predictions. It indicated only that "data significant to the nuclear excavation program were obtained."[96] In other words, if nuclear explosives were buried too deeply, these data had shown, they made mounds instead of craters.

Figure 4.5. With the Sulky mound in the background, unidentified project scientists confer during postshot studies at the Nevada Test Site. *Credit must be given to the University of California, Lawrence Livermore National Laboratory, and the Department of Energy (Atomic Energy Commission) under whose auspices the work was performed when this information or a reproduction of it is used.*

PALANQUIN

Livermore's next nuclear excavation test, Project Palanquin, was planned from the start as a relatively small and private explosion at NTS. Designed as a 4-kiloton explosive, Palanquin was another effort, after the failure of the deep-burial entrapment scheme in Sulky, to create a technical solution to the fallout problem. It was designed to test a new concept—"down-hole" debris entrapment—which, it was hoped, might reduce fallout from earthmoving explosions by a factor of one hundred, making nuclear earthmoving a far more palatable alternative. The idea of the down-hole scheme was that radioactive debris from a cratering explosion could be propelled downward into a cavity below it—for Palanquin, this chamber had been carved the previous year in the Dub device development test—sealing off the new radioactive chamber in the process to prevent venting from other mechanisms associated with the explosion.[97] If successful, it was argued, down-hole entrapment would allow Plowshare scientists to move beyond the slow progress in device development of "clean bombs" and, in turn, allow for larger-scale experiments and demonstration projects demanded (and supported) by the canal study. Palanquin was thus placed ahead of already planned cratering and row-charge tests (see table 4.1), on the one hand, because it might provide useful information for experimental design in later tests, but on the other hand, simply because it was a permissible experiment. If Livermore could not carry out the experimental program that it had originally designed, then the laboratory would settle for one which it could.

Even then, Plowshare scientists had to remain flexible. "We have been continuing our review of the results from the Dub experiment," Chairman Seaborg informed National Security advisor McGeorge Bundy in March 1965, "and their application to prediction of results of the proposed Palanquin experiment. Though we are still confident that the down-hole debris entrapment will work when coupled to cratering, this review indicates that there is sufficient uncertainty as to the degree of its effectiveness that we now plan a more conservative design approach."[98] After increasing the depth of burial from 200 feet to 280 feet—to a scaled depth of burial matching the conservative depth chosen for Sulky—the experimentalists predicted that Palanquin too would probably produce a mound instead of a crater, while still providing a useful test of the down-hole entrapment method for nuclear excavation blasts.

Because a great deal of vented radioactivity was not expected, fallout studies, like those for Sulky, were limited primarily to "close-in" sampling. Also like Sulky, the public information plan for Palanquin was to be tightly controlled. No pretest announcement or still or motion photography was to be released, and all information concerning Palanquin was to be disclosed on an "Official Use Only" basis.[99] As the AEC directed its personnel, compliance with the LTBT meant that the public should know as little as possible about the experiment:

> Every effort will be made to prevent the public from becoming aware that Project Palanquin is scheduled for execution soon. . . . Rather, contacts and responses to inquiry should try to convey the impression of 'business as usual.' In particular, we want to avoid misleading or speculative articles which might create the impression that the U.S. thought it was violating the terms of the nuclear test ban treaty. To

> the contrary, the U.S. does not think it is or it would not have authorized conduct of the experiment.
>
> Unfounded concern [should be minimized] . . . especially during the critical period in which radioactivity may be moving through neighboring states and when the information concerning the radioactivity is OUO [official use only], that might arise among state officials, primarily health officials in Nevada and neighboring states.

A brief public statement was to be released after the event, "describing the crater or mound" and claiming that "data significant to the nuclear excavation program were obtained."[100]

On April 14, 1965, 450 milliseconds after Palanquin was detonated, most of the radioactive gas produced by the explosion burst into the air above ground zero. Entrapment techniques failed completely, as it was later deduced, because of a "stemming failure" of the grout bonding in the emplacement hole. Instead of propelling the radioactive gas downward into the Dub cavity, the high-pressure, high-temperature gas vented like a geyser through the 48-inch-wide emplacement shaft. Gas pressure had likely been magnified by the presence of ten times more water than expected in the explosion. Ten minutes later, the dust cloud from Palanquin ascended eight thousand feet above the earth's surface before "light southerly winds carried the cloud northward into the off-site area." Unexpectedly, the venting created a conical crater, seventy meters wide, twenty-four meters deep, rather than a mound of rubble.[101]

AEC commissioners, project scientists, meteorologists, public health officials, and the president tracked the progress of the dust cloud as it moved northward, but no public statements were issued. Although levels of radioactivity in the cloud were higher than expected, Seaborg told the president that they were still "well below any possible health hazard level even close to the test site." For Seaborg, the primary concern was the movement of what might be interpreted as radioactive debris toward the Canadian border and, with it, a violation of the LTBT. The day after Palanquin, the radioactive air mass was tracked east of Spokane, Washington, and by the following morning it had moved to the vicinity of Butte, Montana. Low levels of radiation were detected at the Canadian border shortly after, "levels so low," Seaborg insisted, that it was "doubtful whether anyone else would detect them."[102]

Public announcements of Palanquin finally came four days after the event. At last, the AEC and Johnson administration put their liberal interpretation of the LTBT to the test: on the basis of internationally sanctioned definitions of "not present" debris, they insisted that the United States had not violated the treaty, and further, that if they *had* violated the LTBT, then they had done so far less egregiously than a recent Soviet test (January 15, 1965) that had transported low levels of radiation to Japan. At the same time, the AEC dismissed fallout concerns within the United States from Palanquin as insignificant. Although low levels of radioiodine were detected in Nevada, Idaho, and as far as nine hundred miles away in Glendive, Montana, it was stressed that concentrations were "substantially below those at which protective actions are recommended by the Federal Radiation Council."[103] But closer to the test site, more proactive observational measures were taken. Six families were taken to the Nevada Test Site's permanent base camp at "Mercury" for whole body counting; four of the families showed "slight thyroid burdens," including eighteen infants and children.[104]

The venting of cavity gas, which had initiated a different erosional mechanism than other hard-rock cratering experiments, had made cratering data from Palanquin useless as data points for scientists' nuclear cratering curves. The unexpected nature of the vent also caused most fallout predictions to be shattered: predicted exposure levels were "low by a factor of 100 at 5 miles, low by a factor of 33 at 10 miles, low by a factor of 10 at 18 miles, and low by approximately a factor of 5 at 40 and 55 miles." Beyond twenty miles from ground zero, it was also noted, "the data were very sparse and the analysis is considered somewhat subjective."[105]

What did Livermore and AEC scientists learn from Palanquin? They learned something about gas-erosional mechanisms in a cratering explosion (a curiosity without value to the excavation program). They also learned, after the Palanquin whole-body counting operation, that a mobile "thyroid monitoring trailer" would be useful for radiation safety operations, so one was acquired by off-site public health and safety personnel. It was later used, in remote areas of the state, to measure thyroid burdens in the field after tests, rather than having to transport families to the test site for counting.[106]

Progress and Resilience

Did the Sedan, Sulky, and Palanquin tests constitute progress in nuclear excavation? All had provided at least some empirical data, contributing to the stocks of data needed for developing cratering curves, models, or computer codes for understanding rock and soil mechanics relative to the physics of explosive design. On the "effects" side of the experiments, the explosions provided data for developing weather-predicting techniques, fallout distribution models, biological-sampling mechanisms, public health and safety operations, and other ongoing, seemingly data-driven scientific work. For the AEC, the representation of these excavation tests—in Official Use Only reports as well as public information releases—thus took shape as an uncomplicated discourse of progress-through-experimentation. Some of this rhetoric was political in the most basic sense, whether as progress reports to funding or regulatory bodies such as the JCAE, AEC advisory boards, or the Interoceanic Canal Study Commission or as technical reports completed by project managers intent on justifying their own experimental designs. But this politics—this ideology of progress—also had a role to play in the everyday work of knowledge production at Livermore (and in the workers' morale), whereby the very generation of experimental data could in itself be justified as progress. As Edward Teller had once observed, "new knowledge in the life of the laboratory—even if that knowledge is disappointing—is most important."[107]

Rather than an uncomplicated story of progress, however, the experimental program in nuclear excavation is better understood as a process of political, geographical, theoretical, and, by all means, pragmatic resilience. Plowshare scientists did *what* they could, *where* they could, to continue to construct experimental knowledge of nuclear excavation techniques, and in this sense, the trajectories of the nuclear earthmoving program—and the conditions under which its knowledge could be produced—were constituted by following the paths of least resistance.

In May 1965, one month after Palanquin, Livermore's Glenn Werth argued, in an updated "planning concept" for Schooner, that the "need for the experiment is self-evident when the comparison is made between the projected applications of excavation . . . and past excavation experience." Still an integral part of the canal study, Schooner was, at this time, in the president's fiscal 1966 budget, and planning for the 100-kiloton cratering blast at the Bruneau River site in Idaho was proceeding apace. By this time, though, the difference between planning an experiment and actually carrying it out were quite obvious for Werth and his colleagues. As Werth saw it, "a program like Schooner is complicated and unfortunately it is the nature of nuclear experiments that planning will change and some uncertainties in authority to execute will exist up to the very end. But the Plowshare concept is a valid one and those in Plowshare believe we are on the threshold of an exciting development. The initiative and spontaneity to get the job done that has served Plowshare and LRL so well in the past will serve the Program again in Schooner." It was precisely the adaptability and resilience of the laboratory, from changes in its experimental designs to its shifting strategies of public information to its accumulated skills in problem solving, that allowed the program to survive. What is more, this resilience was seen not as a weakness or failing of the program but rather, with a slightly quixotic arrogance, as a strength. "When problems arise, as they surely will," concluded Werth, "please contact the Technical Director or his Deputy."[108]

Confined, in a sense, within the borders of the test site, boundaries of classification and authority at NTS were used by Plowshare scientists as elementary resources for knowledge construction in nuclear cratering. At Nevada, Livermore, and other workspaces for Plowshare, scientists made use of the boundaries of classification and those of territorial control so that, while by no means independent of the interests, opposition, and controversies that structured the very nature of the experiments, they managed in these places at least to construct certain facts about nuclear cratering. And it was these facts, as we will see in the next chapter, on which claims about the "technical feasibility" of geographical engineering *elsewhere* were based. But if Plowshare based its claims to truth on these experiments (and related scientific-technical studies), then how could these claims be authenticated, given the nonpublic procedures of observation and self-validation? How could these claims be made to apply to other geographies that Livermore still hoped to engineer? And not least, whose opinions counted in determining what was proper knowledge, and what was not, in regard to questions of public safety?

If Plowshare's grounds for knowledge were in some ways products of the paths of least resistance, then spatial strategies were indeed pivotal to problem solving for Livermore's program builders. Retreats to Nevada, from Idaho as well as from Alaska, and revisions to experiments and to the larger experimental program all depended on territorial variations in legal, scientific, and even moral authority (in the sense that the acceptance of "permissible doses" can be considered potentially a moral as well as a technical issue for us, if not for the subjects of this study). This geography of knowledge construction was by no means incidental; mounds and craters in Nevada, and the absence of such landmarks elsewhere, were at the very root of what could be called progress in nuclear excavation. And yet this constructed progress, such as it was, depended not only on relations inside the laboratory and NTS, and on Plowshare's responses to the external world pressing in upon it, but also, crucially, on "outside" connections forged with common interests in the business, or at least the idea, of nuclear earthmoving. It is to these relationships, and to the drive to take nuclear earthmoving beyond the militarized Nevada desert, that we now turn.

Chapter 5 Pragmatic Engineering Worlds

Feasibility and Trust, Off-Site

"A GOOD NUCLEAR ENGINEER," Edward Teller pronounced in 1969, must also study geography; "how else," Teller asked, "could he engage in geographic engineering, the enterprise intended to change the geography so that it should better serve human purposes?"[1] But perhaps the real question was *how*. The whole excavation enterprise, as the 1968 college textbook *The Constructive Uses of Nuclear Explosives* described it, was best understood from an "engineering standpoint," that is, without considering the "emotional or political issues involved in the use of nuclear explosives. . . . Hence the first step in a nuclear-explosive construction project must be made by the engineer with an adequate supply of information."[2] What a vision, this engineering standpoint! By extricating the political and emotional contexts from nuclear earthmoving, geography was readily grasped as an abstract space-to-be-engineered. Also extricated, from educational texts like this one, from the various professional conferences and symposia where unclassified Plowshare research was presented, and from the engineering manuals produced together with the program, was any sign of the controversies that Livermore and the AEC had navigated through in their pursuit of the experimental program.[3] Instead, experimental knowledge produced from the Nevada Test Site, mulled over and processed through theoretical models and computer codes at Livermore and other sites of Plowshare's classified community, was presented largely as practical, working theory, as engineering technique.

It was no small achievement in rhetoric that the shifting strategies that had guided Plowshare experimentation, and the checkered results of these experiments, could be reconstructed in this manner and, to a degree, and by certain audiences, accepted as convincing experimental knowledge. But these qualifications are important, for as we have seen, claims of Plowshare's progress were not accepted by everyone in the scientific community or in the public at large; for some, the reputation of the AEC and the statements of Edward Teller were by this time quite discredited on matters concerning fallout and radioactivity hazards. But the work of communicating the knowledge of nuclear excavation became a more differentiated social process. When Plowshare was initiated during the late 1950s, its promoters had typically heralded the most spectacular aspects of the program, often directing their rhetoric toward popular audiences in their efforts to generate enthusiasm for the program. But during the years of excavation and underground device development testing in Nevada between 1962 and 1968, the discursive strategies employed for both making and communicating knowledge in the Plowshare program became far more focused. AEC and Livermore scientists now typically directed their experimental discourses toward particular audiences: nuclear physics and engineering societies, potential commercial and government partners in excavation applications both domestic and international, regional booster groups, AEC contractors, and students enrolled in university physics and engineering courses.[4] Training a new generation of civil and nuclear engineers in the techniques of nuclear earthmoving, in this sense, was one means of moving geographical engineering beyond the boundaries of the test site, even if only as a discourse still awaiting its earth-shattering material counterpart.

Given the nature of Plowshare's classified spaces and still limited public information, however, enlisting these actors and institutions in the project of nuclear excavation called for a great deal of trust in the Livermore and AEC scientists on the part of their external audiences: trust that Plowshare proponents were accurately transmitting the results of their work, since the experiments were not witnessed in a public sense; and trust that particular working beliefs about the effects of radiation were true, that is, that low-level radiation below a certain threshold was not cumulatively harmful, even though the so-called threshold theory was discounted by most geneticists, among other scientists. For example, speaking to the JCAE in 1965, DPNE director John Kelly reported progress on "clean device development tests" from the previous year. "The detailed results of these tests are, of course, classified," Kelly stated, "since they provide information on the design of nuclear explosives, but I can say that devices with only a few kilotons of fission can be produced in a wide range of yields up to and including a megaton."[5] Figure 5.1, included in Kelly's presentation, shows the reduction of gamma-radiation exposure from fallout predicted by the AEC for a 100-kiloton test at three developmental stages: the ac-

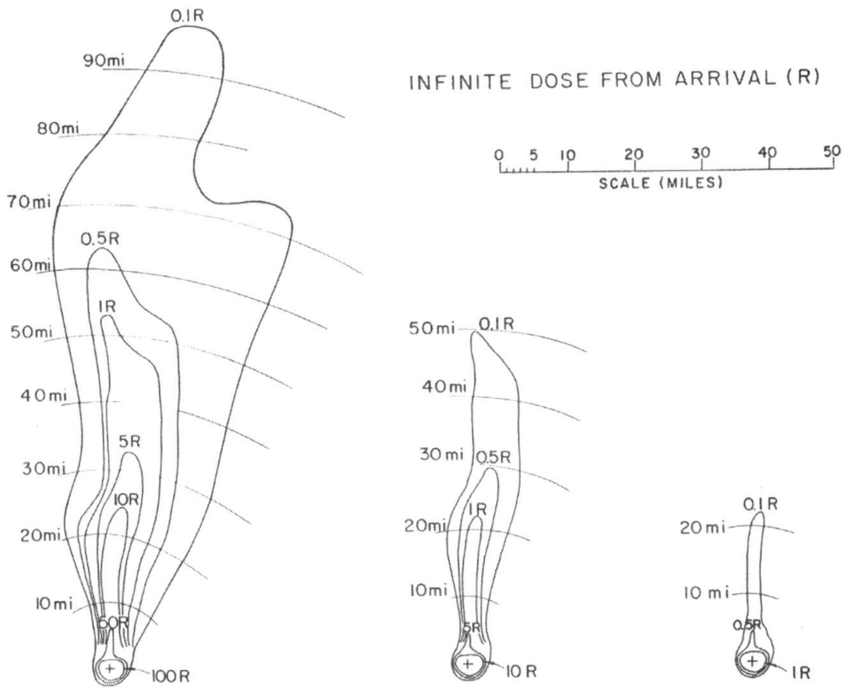

Figure 5.1. The AEC's Division of Peaceful Nuclear Explosions presented these external (gamma) radiation dose estimates for 100-kiloton nuclear cratering explosions in 1962 (left), the "present" (center), and the "future" (right) to Congress in 1965. While the 1962 map is based on data from Sedan, no experimental data were provided in public testimony to support the "present" or "future" estimates. The maps are striking for their complete exclusion of human, biological, or geographical characteristics from the space of the fallout sector. *Source: Statement by John Kelly before the JCAE, January 5, 1965. Plowshare Box P-29, Lawrence Livermore National Laboratory.*

tual infinite doses for external fallout observed from Sedan in 1962, the expected distribution given progress to date in clean-device development, and a further expectation for the future, given continued progress. Because much of the data on which these maps were based were not published, accepting the validity Kelly's claims (and others like them)—as the JCAE did, at least to the extent that funding for excavation research continued in connection with the canal studies (see figure 4.4)—depended to a large extent on trust.

Creating trust, in this sense, was a necessary step as Livermore scientists, and AEC technocrats like Kelly, distilled the discourse of progress-through-experimentation to a more practical technical discourse, coded for application by the engineer with "adequate supplies of information." This discourse was geared partly toward inspiring confidence in the program from its funders and from other sympathetic outsiders who could be brought into the project in some capacity. But it remained always a practical matter as well, as seen in numerous technical "feasibility studies" for potential Plowshare applications that were developed as part of the search for the "useful demonstration project" still considered vital before the Trans-Isthmian canal could be seriously considered. But how much information, and what kinds of information, were necessary to determine whether or not a project was "feasible"? This chapter explores how Plowshare's continuing search for off-site demonstrations of geographical engineering was accompanied by the movement of a particular scientific-technical rationality—embodied in Teller's figure of the "good nuclear engineer"—which took a posture of objectivity by selectively (and subjectively) attempting to exclude the political, scientific, emotional, and moral discourses that threatened to undermine it.

Although it was evident to Plowshare planners that the need to dramatically demonstrate the power and safety of earthmoving could not be met at the Nevada Test Site alone, no matter how many photo spreads *Life* published, it was also clear, after the failure to enact the "model harbor" at Ogotoruk Creek in Alaska, that this demonstration must also be a practical project, an excavation that could be transformed into a useful engineering structure (that is, a canal, road cut, or working harbor). A practical demonstration, as one Livermore physicist put it, would be effective both as a public demonstration of Plowshare techniques and as a useful experiment that would "bridge the gap between rather idealized experiments (such as Sedan, Buggy, and Schooner) and the pragmatic engineering world in which the problems of the environment must be met and solutions found."[6]

In what follows we will examine the mobilization of a discourse of practicality, feasibility, and "safety feasibility" as it moved hand in hand with efforts to enact useful Plowshare excavation projects. How did Plowshare scientists, engineers, and publicists go about "bridging the gap" with the "pragmatic engineering world" through the selection, review, design, and promotion of demonstration projects, and through the cultivation of new technical and political partnerships? Examining the development of would-be demonstration projects in Mississippi, California's Mojave Desert, and Western Australia, along with ongoing developments in the Central American canal studies, I argue that the success of the Plowshare program in enlisting allies in its projects, but ultimately, its failure to blast new experimental spaces in landscapes outside of the Nevada Test Site, were both characterized by, and through, the contested politics of feasibility and trust.

Nuclear Cratering Group

The Plowshare Advisory Committee had recommended in 1963 that it was "desirable to carry out as soon as possible a demonstration project in this country. This could be particularly important in gaining acceptance of the idea prior to conducting Plowshare projects in other countries." But in the wake of Chariot, the PAC had been concerned over precisely which scientists, and indeed, which *sciences*, were actively involved in the preparation of these demonstration projects. The advisory committee "expressed the opinion that the amount of pre-shot ecological surveys necessary to establish the feasibility of a project will vary widely for different sites and cautioned that overemphasis on ecological surveys can adversely affect the Plowshare program."[7] So while the PAC—echoing the sentiments of its Plowshare advisees—recommended that a practical demonstration be carried out quickly, it warned against the co-opting of future projects by ecologists who were perhaps inclined against the establishment of feasibility. As the search for demonstration projects continued throughout the 1960s, Plowshare planners thus sought out collaborators with like interests in the business of earthmoving.

These interests dovetailed smoothly with those of the Army Corps of Engineers, Livermore's senior partner in the Central American canal studies, and with the corps, there was little fear that an ecological agenda would come to dominate a feasibility study. For Lieutenant Colonel Ernest Graves, director of the corps' Nuclear Cratering Group, which was headquartered on the laboratory premises at Livermore, it was an obvious match. As Graves saw it, "LRL has the product and is looking for a market. The Corps of Engineers has a construction job to do and is seeking methods to employ nuclear explosives to do that job. The Corps has the market and is looking for a product." The corps' civil works budget for 1963 included over eight hundred million dollars in construction costs, much of which was devoted to earthmoving and rock excavation. But while Graves and his colleagues were indeed eager to use nuclear explosives for such tasks, it was also clear to them that the endeavor faced serious challenges. The development of nuclear excavation technology, Graves surmised, was hampered by two rather paradoxical problems:

> One is that nuclear excavation promises the greatest advantages in big projects, yet we cannot expect approval for these big projects until the technique has been proven by actual experience with smaller projects. The other is that less developed regions outside the United States offer more potential for nuclear excavation projects than our own highly developed country, yet we cannot expect other countries

> to accept this new technique until we have proved it by actual experience at home. Thus, our first practical nuclear excavation projects are likely to be some of the most difficult technically.

"However difficult," Graves concluded, "we can certainly solve the technical problems."[8]

The work of the Nuclear Cratering Group followed three principal lines: (1) experiments with chemical high explosives, including "row-charge" experiments to extend studies dealing with canal construction (carried out in conjunction with the Sandia laboratory); (2) studies of the engineering properties of nuclear craters; and (3) feasibility studies for the use of nuclear explosives in corps civil works projects. Reflecting the pragmatic mission that Graves envisioned for the Nuclear Cratering Group—with one foot in the laboratory, the other "on the job" with the corps—the development of an engineering manual for nuclear excavation in the spring of 1963 was among the Nuclear Cratering Group's first projects. Its purpose was "to provide engineers with technical information and guidance which can be used to study the feasibility of executing large-scale earthmoving projects with nuclear explosives." It was recognized that nuclear excavation techniques were still under development, and the manual was thus subject to change, but in the meantime it could be used to translate the principles of nuclear cratering and other techniques to the engineers assigned to the group's new office at Livermore and to corps district offices throughout the country. The manual included introductory models of nuclear excavation, an outline of nuclear explosives data required for executing feasibility studies (for example, cratering curves, radioactivity estimates, explosive emplacement techniques, and associated costs), and discussions of slope stability and other engineering properties of nuclear cratering in different soil media. A separate section on "safety considerations" dealt with radiation hazards. Because of the complexities associated with ingested internal hazards, however, no evaluation of the radioiodine problem was presented, indicating instead that, for projects that appeared feasible in every other way, the question of internal hazards would be evaluated by AEC specialists. Otherwise, the engineers were informed that "the use of present fallout prediction techniques with intelligent allowance for variations has proven adequate to ensure public safety."[9]

The Nuclear Cratering Group's manual thus codified the state of the art not for theorists but for future practitioners of nuclear excavation. At the same time, one of the reasons for working out the technical how-to guide before nuclear excavation had been put to use as an applied scientific technique was so that solutions to some of the engineering problems that might arise could feed back into the research and development process: "One of the objectives of this manual," according to its introduction, was "to stimulate the exchange of ideas between the scientists and engineers who are developing nuclear excavation and the engineers who will be the main users of this new method of construction."[10]

TENNESSEE-TOMBIGBEE WATERWAY PROJECT

The Nuclear Cratering Group advanced several ideas for practical nuclear excavation projects within the United States. Group site-selection procedures differed from those that had led to Livermore's choice of Ogotoruk Creek for a nuclear harbor, where it had essentially looked for certain physical and locational conditions first, and worked backward to the economic rationale for the project. The corps instead looked at existing excavation schemes already under consideration for conventional blasting, including plans for the construction of dams, canals, and other aspects of river system management, as sources for developing its feasibility studies for nuclear earthmoving jobs. The most comprehensive feasibility study developed by the corps and its Nuclear Cratering Group (with the exception of the Central American canal study, as we shall see) concerned a canal project known as the Tennessee-Tombigbee Waterway. First considered by the corps as early as 1874, the scheme called for a 253-mile-long waterway excavation in northern Mississippi, along with the construction of a series of locks and dams, that would connect the Tennessee River to the Tombigbee River, which empties into the Gulf of Mexico at Mobile, Alabama (figure 5.2). The new waterway, it was argued by boosters from the Mississippi Industrial Research Center, the Tennessee-Tombigbee Waterway Development Authority, and the Corps of Engineers Mobile District, would facilitate commerce in the region by connecting it to the mid-American waterway system. In northern Mississippi and Alabama, where a burgeoning chemical-manufacturing industry was taking root, proponents argued that the "opening of this waterway link with its accompanying transportation facilities will serve as a major attraction to industry bringing greater employment and increased income to cities and states along its banks."[11] Standing in the way of such visions was a wooded hill, rising 560 feet above sea level, which separated the Tennessee and Tombigbee rivers.

As early as 1960 the Development Authority, which had been sanctioned by the state governments of Alabama, Mississippi, and Tennessee to promote the project, was arguing that hydrogen bombs "could be used to good advantage" in canal construction, a contention that it tied explicitly to the "great strides . . . made in the perfecting of clean nuclear devices."[12] Evidently believing the proclamations of Teller and others concerning the imminent safety of PNEs, in 1961 Mississippi governor Ross Barnett called for "bold and

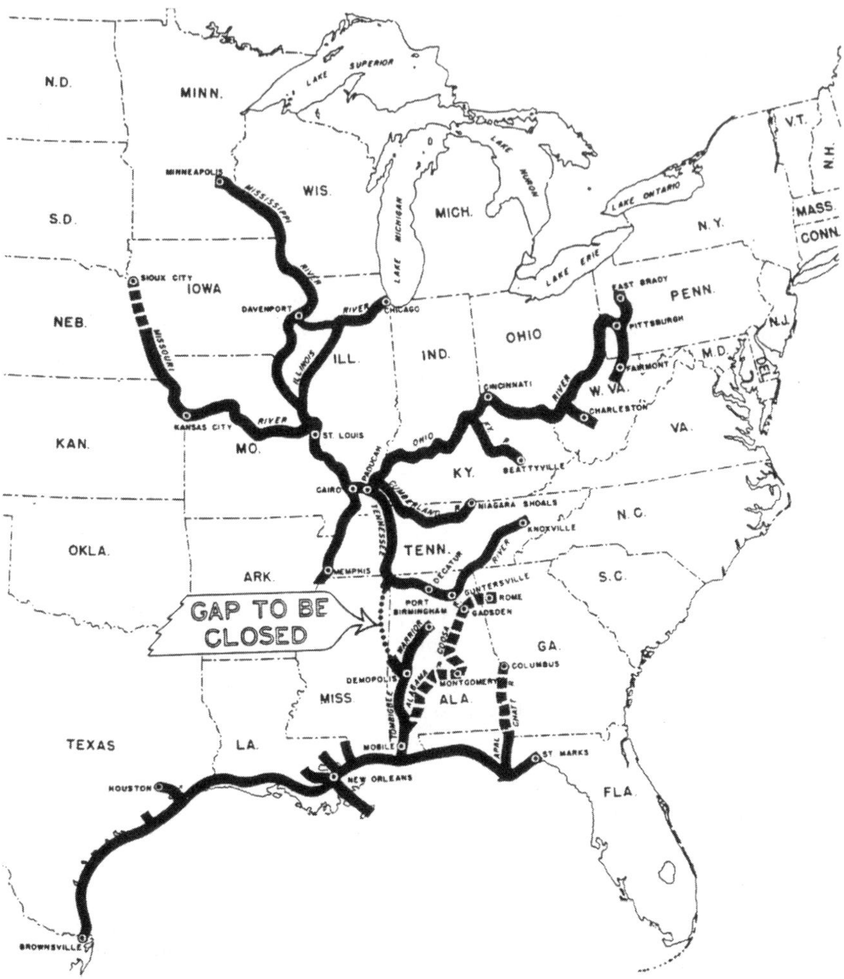

Figure 5.2. Nuclear excavation was proposed by the U.S. Army Corps of Engineers, Mobile District, as a method for cutting portions of the "connecting link" between the Tennessee and Tombigbee rivers, as shown in this Tennessee-Tombigbee Waterway Development Authority brochure (no date), Plowshare Box P-7, Lawrence Livermore National Laboratory. The case for the new waterway is strengthened by the absence of the Mississippi River on the map.

immediate action" and announced that he would "recommend to the Atomic Energy Commission the use of 'clean nuclear explosives' in making the crucial 25-mile canal cut that would connect the Tennessee and Tombigbee rivers."[13] A feasibility study of the nuclear excavation project was ultimately prepared by the corps in 1964. As the Development Authority lobbied for support, it was eager to stress for Commissioner Seaborg the practicality of the proposed project, noting that "the Corps of Engineers have found the Tennessee-Tombigbee Waterway project to be highly favorable and it seems most practical that these initial experiments be made on a practical project which would be made even more attractive."[14]

The next governor of Mississippi, Paul Johnson, was equally anxious to open his state to geographical engineering. He was also confident that the largely impoverished residents in the area of the proposed blasts, who had done little to prevent the arrival of the chemicals industry, would not stand in the way of the waterway project. "The people of the State of Mississippi," he assured Defense Secretary Robert McNamara, perhaps mistaking Plowshare as a province of the Defense Department, "will . . . accept the proposed construction method with an open mind and if the study proves feasible there will be very little, if any, organized opposition."[15] As "supporting evidence," Johnson pointed to "excellent relations" between the AEC and the people of south Mississippi, where the AEC was planning to detonate a 5-kiloton underground explosion later in 1964 as part of Project Dribble. Dribble, which was not part of Plowshare, consisted of two separate nuclear detonations in an underground salt dome four miles northeast of Baxterville, Mississippi. The first detonation, a 5.3-kiloton explosion code-named Project Salmon that was carried out in October 1964, created a cavity in the salt formation 110 feet in diameter, while the second, code-named Sterling (conducted over two years later) was a .38-kiloton explosion inside the cavity that was intended to gauge how the shock waves generated in such an underground explosions could be masked inside salt formations, and thus to improve detection and verification techniques related to the monitoring of Soviet underground tests.[16] Governor Johnson was eager to bring peaceful nuclear explosions to his constituents. As he explained to Seaborg, "I can assure you that the citizens of the State of Mississippi will look on further experimentation within the State with the same constructive and unemotional attitude they have shown toward Project DRIBBLE and that evaluation of nuclear explosives in the construction of civil works will be welcomed rather than opposed."[17]

A PROBLEM OF PHILOSOPHY?

However favorable the corps may have found the nuclear Tennessee-Tombigbee Waterway, and however trusting were Mississippi governors and industry boosters that Livermore's explosives were sufficiently "clean," scientists and engineers in the classified community—those with actual knowledge of the state of clean-device development—turned out to be far less sanguine about the project. According to a study made by the corps' Mobile District, nearly 1,500 people lived within five miles of the proposed blast area, and over 340,000 people lived within fifty miles.[18] As one Sandia engineer who reviewed the project put it, "there are so very many people nearby."[19] And the design plan called for so very many explosions. As the idea was worked out by engineers from the Mobile District, the Nuclear Cratering Group, and a subcontracted Houston engineering firm, nuclear explosives would be used to excavate a 5.3-mile-long channel, a section in northeast Mississippi that constituted the key piece of earthmoving necessary for the project. The problem was to excavate a 300-foot-wide channel, with depths ranging from 65 to 115 feet. Because of layers of clay below the surface, however, it would be necessary to "over-excavate" somewhat in order to compensate for slope stability problems. To make this cut, a configuration of eighty-one nuclear explosives was necessary, ranging from 10 to 50 kilotons in explosive yield, for a total of 1.9 megatons.[20]

Perhaps because Tennessee-Tombigbee was the first project under consideration that did not originate within Livermore's experimental program, a new "inter-agency organization" was developed to guide evaluative procedures. Although the project was placed under the direction of the corps' Mobile District and Nuclear Cratering Group, it was subject to Livermore review at several stages, while the AEC's Nevada Operations Office was responsible for the project's safety program.[21] These reviews would eventually doom the nuclear waterway proposal long before Governor Johnson's predictions about his state's acquiescent polity could be put to the test. As Livermore's Gerald Johnson put it bluntly, the laboratory did not feel, at least as of January 1965, that the project was worth pursuing. The key problem was that, because of expected levels of radioactivity, a fallout sector sixty to seventy degrees wide, to a range of forty to fifty miles, would need to be completely evacuated of people for an unspecified time. Even assuming that the radioactivity problem could be reduced to an "acceptable level" through advances in clean-device development, Livermore studies suggested that the "magnitude of the seismic hazard alone would appear to make the use of nuclear explosives to blast the channel discouragingly expensive."[22]

The review by the AEC's Nevada Operations Office more explicitly assessed the hazards of the project. Evaluating a study made jointly by the Nuclear Cratering Group and the Mississippi Research and Development Center, the AEC was compelled to shed light on several key safety factors that had not been addressed. For example—as per the instructions in the Nuclear Cratering Group's excavation manual—the I-131 problem had not been examined at all, but would be a serious concern, the AEC now stated, in spaces up to two hundred miles downwind from the blasts. The soil media and groundwater could be expected to present problems as well: first, though little was known of the phenomenology of nuclear explosions in limestone, it was obvious that detonating in such a water-saturated media would dramatically increase fallout levels; and second, the explosions would contaminate aquifers serving parts of Mississippi and Alabama. Operational safety costs, the AEC's safety review estimated, would exceed twenty-five million dollars, along with perhaps another ten million for long term, postshot groundwater studies. Finally, however confident was the governor in the "constructive" and "unemotional" attitudes of area residents, the AEC's prior experience led it to a more skeptical view:

> An intensive, prolonged, and effective public information and public education program would be required to avoid regional and even national dissatisfactions with handling of the entire nuclear program. As an illustration, residents might be quite unwilling to return to their homes if the area has been contaminated by fallout, although this has not been a problem in Nevada. The large numbers of people involved in numerous evacuations, including those who would require special handling such as the new-born, the ill and the aged, would add to the general difficult of the public information and community relations problem.[23]

The AEC would not rule out "safety feasibility" for Tennessee-Tombigbee—to do so would have reflected badly on other large projects still under consideration, not least the Central American canal study—but it did not recommend that any further action be taken on the project.

Lieutenant Colonel Walter Slazak, the second director of the Nuclear Cratering Group, did not hide his disappointment over the AEC's appraisal of the Tennessee-Tombigbee project. He responded to Nevada operations manager James Reeves, expressing concern about how the AEC's evaluation of the corps' feasibility study would affect the entire Plowshare program. For Slazak, the problem was one of philosophy:

> As you know, I have repeatedly objected to the philosophy that for a feasibility study one should assume worst possible conditions and then design a program and cost estimate consistent with such pessimistic assumptions. This philosophy can kill any potential nuclear excavation project and in turn destroy nuclear excavation without a fair trial. I appreciate your grave concern for safety in the conduct of a nuclear event and the areas of uncertainty which exist under the present technology. However, I feel it is necessary and appropriate to project ourselves in time to when nuclear excavation has become truly operational rather than to assume the technology will always remain in the present research and development status.[24]

Slazak feared that, without projecting in time to a more advanced state of the art, Plowshare risked being consigned to only the most marginal places. "The feasibility of applying nuclear excavation to the Tennessee-Tombigbee divide cut is not the issue here," Slazak argued. "The important issue is the implication that projects of this size are categorically infeasible unless located in the most remote areas of the world." And so he requested that Reeves produce an additional appraisal, "assuming more desirable conditions."[25] The Nuclear Cratering Group was still sponsoring eight additional studies from other corps districts, and he feared that the AEC's pessimistic philosophy might jeopardize the whole endeavor. Reeves, however, did not comply with the engineer's request, responding by insisting that the project "has rather severe safety problems and our statements were aimed specifically at assessing these situations without attempting to draw any conclusions to general applicability."[26]

Even so, with its own pragmatic, optimistic philosophy, the Nuclear Cratering Group had inadvertently traced a new limit to geographical engineering: Slazak was right. Large nuclear excavation projects *were* categorically infeasible in all but the most remote areas (even as remoteness itself, for Livermore's nuclear scientists, had proved to be a contested issue). The real mistake—committed by everyone from Corps district engineers to state governors to Development Authority boosters who had pursued the project—was that they had believed the public claims about progress in clean-explosives development; they had trusted Plowshare too much. Undeterred, for its part, the Nuclear Cratering Group carried on its work in support of nuclear excavation throughout the decade.

Planetary Ambitions

The shift into the somewhat uncomfortable role of critics of a nuclear excavation project, on the part of the Livermore and AEC scientists and engineers, was not the radical reversal of standpoints that it may seem; the deliberative proving grounds of Plowshare's internal review had found the public safety problems expected from the Tennessee-Tombigbee project simply too severe, by their standards, to warrant serious consideration from the Plowshare program. Many other projects, however, were considered more viable.

The search for potential excavation sites continued not just nationally but globally, through the widespread cultivation of friendships and enlistment of allies in nuclear earthmoving technology. Ideas for practical earthmoving projects, some solicited, others volunteered, poured into Livermore from a wide range of sources. Some came from direct institutional linkages to the laboratory, like the engineer Corps' district offices and the U.S. Geological Survey, but others were generated from less public relationships with contract engineering firms such as the Hudson Institute of Harmon-on-Hudson, New York. This latter firm, which listed a former Colombian minister of public works as a consultant, investigated on Livermore's behalf several "'high leverage' civil engineering projects in Colombia and elsewhere in Latin America."[27] Without any publicity, the Hudson Institute conducted preliminary feasibility studies, with the cooperation of the Colombian Air Force, of two possible flood control projects in Colombia. Other ideas were sent to the laboratory by a variety of politicians, development professionals, and amateurs whose geographical imaginations had been stimulated by the Plowshare idea. One proposal on file at Livermore, for instance, originated from the secretary of the Canal Association of Milton, Florida, who informed the Plowshare director that his organization had the "necessary authority" to dig a channel near Pensacola.[28] More seriously considered at Livermore was a proposal from a quixotic water resources engineer from the US AID mission to Ghana, suggesting a water diversion from the Niger River basin into the Upper Volta River basin.[29]

The most promising projects, however, seemed to be those pursued through commercial and government contacts, at home and abroad, through senior Livermore scientists, including Edward Teller, still Plowshare's most effervescent ambassador. One Nat Harrison, for example, invited Teller to Quintana Roo on Mexico's Yucatán Peninsula for a "quick look-see of the nuclear explosive possibilities related to their greatly needed harbor facilities," and enclosed for Teller a carbon copy of the letter he had written to the governor of Quintana Roo describing the project idea. "The climate is wonderful," Harrison added cordially, and "the London banker's home with complete staff I think you will find most enjoyable and comfortable."[30] Still more coveted as a site for geographical engineering, however, was Australia. As Teller described it in his 1962 *Legacy of Hiroshima:*

> In Australia, too much water runs the wrong way. Most rivers run down the eastern slope of the mountains near the Pacific, ending their short courses in the ocean and leaving Australia's western reaches a desert. The streams could be diverted to irrigate the desert. We can pierce the mountain ranges and excavate a system of necessary reservoirs most easily and most efficiently with nuclear explosives. The savings over conventional methods of moving large quantities of earth would be enormous. This project is a dream only because it is so great. Courage and enterprise could make it a reality.[31]

Contacts with Australians, and the development of a small cadre of Plowshare experts in the Australian Atomic Energy Commission, were initiated as early as 1962 in connection with several prospective geographical engineering projects.[32] Livermore's Gary Higgins made a "Plowshare tour" of the country in 1962, and Teller visited in 1966. His tour contributed to the development of an international agreement to formally investigate a nuclear harbor project for the state of Western Australia, as we shall see.

Table 5.1 catalogs the numerous nuclear excavation projects on file at Livermore under the Plowshare program. Since the level of consideration devoted to these potential applications was highly variable, ranging from a mere exchange of letters to preliminary "paper studies" to formal technical feasibility studies, the purpose of the table is to present the technical and geographical breadth—and not the depth—of the projects that Livermore scientists spent some of their time thinking about, rather than to suggest that each of these was actively pursued. The scale of the projects varied as well, ranging from the continental, including the excavation of a new superhighway traversing the Andes and the western Amazon, to highly localized modifications of existing commercial and transportation facilitates, such as the clearing of rocks that impeded navigation or the excavation of a harbor to facilitate the extraction of a particular ore deposit. While the regional contexts in which these engineering megaprojects functioned were quite diverse, Plowshare scientists seem to have had little trouble enlisting at least some allies in the business of earthmoving, radiation controversies aside. Nor was it especially difficult to graft the rhetoric of geographical engineering onto existing logics of international modernization and development. Nearly all of the projects listed here were geared in some manner toward the incorporation of more remote resources into global networks of production and exchange and toward smoothing out the physical conditions of planetary commerce.

But outside enthusiasm for nuclear earthmoving was not without its own practical limits. Whereas the right group of engineers might have been willing to take on any one of these projects, using the "best available" techniques and meeting environmental, health, and safety problems more or less as they arose, the potential hosts of geographical engineering projects were more apt to raise questions about the excavation enterprise. Some were apprehensive about being the *first* demonstration project. Was it an experiment, or was it a working technology? Although Plowshare scientists frequently showed off their Nevada craters to prospective earthmovers and enthusiastic journalists, questions about Chariot's cancellation, and the failure to enact a practical excavation project elsewhere, dogged the program. For example, one Peruvian supporter of the nuclear excavation of road cuts for a new trans-Andean highway reported to Gerald Johnson that "the attempt at getting Plowshare information to the President was not successful. His cousin, Dr. Muñoz, refuses to believe your nuclear devices are beyond the theoretical stage. After speaking with him, an attempt was made at the Highway Commission. They asked, 'why don't they use this method in the States, if it is so good?' The discussion that followed was very negative."[33] Similarly, in Western Australia one avid Plowshare supporter pointed out in 1965 that "the subject of Project Chariot is sure to be raised and the question asked, 'if Chariot has not gone forward . . . why hasn't it?' You must admit this is a curly question and one which must be asked or people who are the enemies of both our countries will very quickly ask it for us."[34]

Livermore scientists and AEC planners did not need these questions posed to them from the other side of the world; they had long been aware of the need to demonstrate nuclear excavation within U.S. territory before the technology could be delivered to others. The first demonstration project developed after Chariot had been incorporated into the five-year nuclear excavation plan for precisely these reasons. Project Carryall was designed as a practical demonstration of nuclear earthmoving much closer to home for the Livermore scientists, and was considered a far more feasible project than the corps' Tennessee-Tombigbee Waterway.

Demonstrating Feasibility at Home: Project Carryall

In June 1963, two San Bernadino, California, newspapers broke the story that "mankind's first civilian atomic explosion" might be used to blast a new highway and railroad pass through the Bristol Mountains in California's Mojave Desert. Both papers were owned by James Guthrie, a member of the California State Highway Commission.[35] The highway commission, it was reported, was considering a re-alignment of the storied Route 66 between Needles and Barstow, and if approved, nuclear explosives could be used to shorten the route considerably by cutting "a swath 200-to-400 feet-deep through an almost solid rock mountain."[36] The newspapers had learned that state highway engineers, along with several officials from the Atchison, Topeka, and Sante Fe Railway—who were considering a similar re-alignment to their line—had recently visited the AEC's Nevada Test Site, where they had examined for themselves the cratering characteristics of nuclear explosives. The leaked reports concluded, inevitably, with quotes from Teller, here indicating that the day was near when nuclear earthmoving could be accomplished so safely that one could walk inside the crater, unharmed by radiation, the very same day that it was excavated. In the meantime, as the people of southern California's "inland empire" were informed, highway and railroad engineers were hard at work attempting to "determine the feasibility of atomic explosions in civilian highway and railroad projects."[37]

Table 5.1
Nuclear Excavation Projects Considered under Project Plowshare by Type, 1957–1973

Type of structure operation	*Location of project by country or region*
Canals: Navigation	Paraná River, Argentina Medeira River, Brazil Mackenzie River Delta, Canada Simpson Strait, Canada Atlantic-Pacific Interoceanic Canal: Central America (routes through Panama, Colombia, Nicaragua) La Araracuara, Colombia Elat-Dead Sea Canal, Israel Mediterranean to Red Sea, Israel Luzon Island Canal, Philippines Seoul to Yellow Sea, South Korea Isthmus of Kra, Malay Peninsula, Thailand Tennessee-Tombigbee Waterway, Mississippi, USA
Canals: Water diversion	Lake Titicaca, Bolivia Tempisque Valley, Costa Rica Mediterranean to Qattara Depression, Egypt Ganges to Hooghly, India Andes Mountains, South America (water diversion: east slope to west slope) Jonglei Diversion Canal, Sudan Mediterranean to Chotts Depression, Tunisia and Algeria Arizona, USA Yukon River Diversion, Alaska, USA Trinity, Mad, and Eel Rivers, California, USA Niger-Upper Volta River Diversion Canal, West Africa
Harbors	Cape Keraudren, Australia Arica, Chile Salaverry, Peru Cape Thompson, Alaska, USA Kuanakakai Harbor, Molokai Island, Hawaii, USA Nome, Alaska, USA Prudhoe Bay (offshore "atoll"), Alaska, USA Shenya Island, Alaska, USA
Overburden removal	Erzberg Project, Greenland (Denmark) Madagascar
Railroad and highway cuts	Chile to Argentina (three routes) Buenaventura to Bogota, Colombia Trans-Andean Highway, South America Boca Pass, California, USA Bristol Mountains, California, USA
Water resources: Dams	Diamond Gorge, Australia Río Bío Bío, Chile Nari and Hab Rivers, Pakistan Tarbela Dam, Indus River, Pakistan

(continued)

Table 5.1 Nuclear Excavation Projects Considered under Project Plowshare by Type, 1957–1973 *(continued)*

Type of structure operation	*Location of project by country or region*
	Camelback Dam, Gila River, Arizona, USA Cochiti Dam, New Mexico, USA Rampart Dam, Alaska, USA Twin Falls, Idaho, USA
Water resources: Reservoirs	Paraguay River, Argentina and Paraguay Ord River, Australia Beni River, Bolivia North American Water and Power (NAWAPA) project, Canada and USA La Macarena, Colombia Ambukalo Reservoir (silting basin), Philippines Mekong River, Southeast Asia Swamp Drainage, South Korea Feather River Project, California, USA

SOURCE: The classification scheme for this table, along with much of the data, is based on L. J. Vortman, "Nuclear Excavation," SC-DC-69-1716, March 1969, Sandia Laboratories. Plowshare Box P-19, LLNL. I have augmented Vortman's list by adding to it additional projects for which I found at least some record in the Plowshare Collection, Office of History and Historical Records, University of California Lawrence Livermore National Laboratory.

The excavation of the Bristol Mountain pass, located fifteen miles south of Amboy, California, was intended to produce a number of practical benefits. The state Division of Highways had been planning to change the path of U.S. 66 as part of the construction of new Interstate Highway 40, and was considering two possible routes: the nuclear-blasted swath through the Bristol Mountains, and another route just one mile to the north. Project Carryall, developed jointly by the Division of Highways, the railway, and Plowshare, brought the highway re-alignment together with one contemplated for the railroad. The present rail line, it was argued, did not follow the shortest route, but rather one that had been the "most economically feasible at the time it was constructed in 1883." Carryall would produce a better grade across the Mojave for the Santa Fe line and shorten its route by fifteen miles, reducing travel time by about fifty minutes and thus lowering operating costs. The railway had recently completed a similar change—sans thermonuclear bombs—to its freight line in northern Arizona in 1960, where a twenty-six-million-dollar relocation saved about one hour per trip. Without the joint effort with Plowshare and the Division of Highways, however, the new line across the Mojave would require construction of an expensive two-mile-long tunnel, putting the costs of the project out of reach.[38]

The rapid growth of industry in California had placed "extreme demands" on existing transportation facilities. The railroad and trucking systems were "virtually connecting segments of vast transcontinental assembly lines which move goods and materials to waiting manufacturers and consumers."[39] For Livermore, Carryall seemed an ideal demonstration project. Not only would the road cut be converted into a "useful engineering structure," it was also located in what Plowshare scientists saw as a sparsely populated desert region, reducing the sorts of complicated and economically untenable evacuations that safety reviewers for the Tennessee-Tombigbee had complained about. In addition, nearly the entire project area was owned by the state of California, the U.S. government, and the Southern Pacific Railway, limiting possible challenges over land tenure issues. Since the AEC would subsidize the project by paying for the excavation costs (including the related public safety program), the railway and the state saw Carryall as an attractive prospect for the reroutings, and the parties agreed to collaborate on a technical and economic feasibility study of the project.

How was the nuclear solution feasible? Livermore's design for Carryall described the carving of the two-mile passage with twenty-three explosives, detonated in two phases, with a combined yield of over 1.8 megatons of energy.[40] According to the feasibility study, the explosions were expected to release about 660 tons of radioactive fission products.[41] In other words, although Carryall was to be eighteen times larger than Sedan, Livermore scientists claimed that it would produce about two-thirds as much fallout—a projection that assumed remarkable progress in clean device development. With highway and railroad construction timetables depending on it, Carryall was to be converted into a working structure as quickly as possible, which meant that these assessments of radioactivity were then brought to bear on the question of when workers could safely re-enter the area to build the four-lane highway and double-track railroad. One Livermore draft report indicated that all on-site workers, up to six months after the blasts, were expected to "encounter some airborne radioactivity[,] and respiratory equipment will be required." Protective clothing would be mandatory for up to roughly a year.[42] If workers were thus outfitted, it was later argued, forty-hour workweeks would be possible, with "proper radiation protection precautions," within about four days of the blasts.[43]

Ultimately the feasibility study drew four conclusions: (1) the project was technically feasible, but to minimize safety problems, projects Buggy, Schooner, and Galley should be conducted in advance (see table 4.1); (2) the projected timetable for Carryall was compatible with the needs of the highway commission and the Sante Fe; (3) the cost of the nuclear solution was estimated to be roughly eight million dollars less than by conventional methods, *exclusive* of nuclear explosives and safety costs; and (4) further studies were needed related to safety considerations, drill holes, seismic conditions, hydrology, population, and weather.[44] The AEC released a summary of these principal findings to the public in December 1963, along with a general description of Carryall. The press release specified the preliminary nature of the feasibility study, but nevertheless expressed confidence: "It appears at this time that, owing to the remote location of the site, protection of the public from harmful radiation can be assured."[45]

REVIEW

The feasibility study was passed to the AEC's Nevada Operations Office for further review in early 1964. This review, in which comments on the feasibility study were solicited from Public Health Service, Weather Bureau, and Sandia laboratory scientists and engineers, was to cover all aspects of Carryall, but focused primarily on areas related to the public safety program, which was AEC Nevada's chief responsibility. In this regard, Sandia's M. L. Merritt was assigned the task of making an "independent analysis" of the safety of Carryall.[46]

"The Carryall feasibility report," Merritt noted at the beginning of his review, "includes nearly every safety factor that should be considered except the offsite effects of the dust cloud." Compounding this problem, Sandia's estimates of radioactivity release turned out to be far greater than Livermore's, owing to "greater conservatism and in different methods of extrapolation from existing data." In fact, Sandia's radiation-dose rate predictions were five times higher than Livermore's, and extended twice as far in distance. Although such discrepancies were quite common, Merritt noted, and "indicative of basic uncertainties in the field," the Sandia estimates made Carryall "a more marginal proposal than its feasibility statement implies." Radioiodine was a primary concern for Merritt, especially since some year-round dairying was conducted in the valley around the city of Needles, California, just sixty-eight miles east of the proposed road cut in the direction of the preferred fallout sector. The location of Carryall was certainly more remote from population than, say, the Tennessee-Tombigbee project, but there were still thirty thousand people living within seventy miles of ground zero. Ultimately these safety considerations caused Merritt and his Sandia colleagues "not to be as enthusiastic about the feasibility of Carryall as the authors of the feasibility report." And yet Merritt did not rule Carryall out. "Nevertheless," he continued, "under certain conditions the project can probably be carried out safely."[47] At least two settlements—Danby and Essex, California, with populations of ten and one hundred, respectively—would have to be evacuated, and Merritt also recommended additional safety measures: residents of other nearby towns should be film-badged (a common technique for measuring external radiation from tests); traffic should be stopped on highways between Amboy and Kingman, Arizona; structures within fifteen miles should be surveyed for possible seismic and air blast damage; groundwater surveys should be conducted around the project site.

Scientists from the Public Health Service and Weather Bureau largely concurred with Merritt's evaluation. Stressing that Merritt's radioactivity estimates, rather than Livermore's, should be used for planning purposes, one meteorologist added that dairying activities might need to be surveyed in parts of Arizona and New Mexico as well, because of the possibility of radioactive hotspots after Carryall.[48] Entirely absent from the feasibility study, as the AEC's John Philip observed, was any bioenvironmental program; yet, notwithstanding the PAC's systematic devaluation of ecology for establishing the feasibility of demonstration projects, Philip indicated to Livermore's Gerald Johnson that "some attention will be needed in this area if the project is to go forward."[49] When, in October 1964, Livermore advanced a preliminary plan for Carryall, it therefore included a highly circumscribed "biological evaluation program," designed to assess the biological impact of the blasts on the environment, particularly the "determination of the radionucleides entering man's food." Following the approach adopted for Schooner in Idaho, and decidedly contra Chariot, the biological program was to be conducted only by "appropriate agencies" experienced in working with the AEC.[50]

Despite the ambiguous interagency review, the project was authorized to move forward, with detonation planned for 1967 (contingent on the successful completion of the Buggy, Schooner, and Galley tests before it).[51] Meanwhile, as engineering surveys began at the project site in January 1965, the AEC hoped to fly more or less below the public's radar, precluding even the positive endorsements of the supportive San Bernadino newspapers. As the AEC public information specialist (and Chariot veteran) Rod Southwick put it: "We expect this to be a low-key program, conducted in many instances in an informal manner to avoid the trappings of any high powered sell job."[52]

WHITHER CARRYALL? END OF THE LINE

When Livermore sought to enact Chariot in Alaska, the demonstration project had depended on little else than the evaluation of that project with the notable exception of the test moratorium until mid-1961. Of course the procedures for determining the preshot evaluation of Chariot were complex and highly contested; even turning the Chariot bioenvironmental program into an evaluative process, rather than simply a comparative pre- and postshot biological experiment, was the product of a great deal of oppositional science and politics. Now, just a few years later, the second demonstration project that Livermore actively pursued depended on much more. It depended on establishing a particular interpretation of the LTBT, or its amendment; it depended on geographically expansive networks of surveillance of milk cows and other ecological considerations; it depended on a great deal of progress in nuclear cratering experimentation in Nevada. And in this last regard, as we saw in chapter 4, Plowshare continued to experience numerous delays and setbacks.

In June 1965, shortly after Palanquin had taught project scientists one way not to design radioactivity entrapment techniques, Gerald Johnson was compelled to write John Kelly about the status of Carryall: "As you are aware, limitations imposed by the present interpretation of the Limited Nuclear Test Ban Treaty have forced postponement of the Schooner event pending development of nuclear explosives or emplacement techniques capable of reducing venting radioactivity by a significant factor over presently available designs. . . . Because of the intimate relationship between the nuclear excavation program . . . and the Carryall excavation, it has been necessary to delay the Carryall schedule by approximately 18 months."[53] Johnson noted the difficulty of setting any schedule at all, given the uncertainties of the program, but suggested optimistically the fall of 1968 for the Carryall detonations.

This delay, as Johnson was aware, would not put Carryall in good stead with the highway commission, which wanted the road cut for the new Interstate 40 to be blasted as soon as possible. He was also aware that delays in the experimental program jeopardized more than just the Carryall demonstration itself. As the PAC stressed at a November 1965 Las Vegas meeting, it was essential "that an adequate series of excavation experiments culminating in a Carryall-type project be conducted in the U.S. in the reasonably near future."[54] The failure to do so, the PAC believed, reflecting the concerns of many program supporters, would hamper even the evaluation of nuclear excavation for the Central American canal study.

Poor Plowshare. For Livermore scientists, these delays always seemed to be caused by outside "political" forces; nuclear cratering was forever held back by matters of politics that they believed had little to do with the actual science of it. In this sense, Johnson and others refused to draw connections between the political limits imposed on Plowshare (and nuclear testing in general) and the dangers widely attributed to fallout from testing by genetic and biological scientists. But no matter how unnecessary the nuclear experimentalists may have felt these delays in testing were, Carryall was—unlike Chariot before it—already at the end of the line in the experimental program, and the project could not be pushed forward until more of the preliminary experiments were performed. As such, Carryall was not formally canceled, but after the initial postponement requested by Johnson, the project vanished from all reports to Congress and other AEC public discourses. With Buggy and Schooner delayed, as it turned out, until 1968, the highway commission turned to conventional methods for its re-alignment. The standard, explanation for the failure to enact Carryall that later emerged was that, though the project had been found to be technically feasible, nuclear excavation "could not be accomplished on the time scale required for completion of that section of the interstate highway system."[55] Whatever the case, Plowshare would have to carry on without it.

"Safety Feasibility" and the Trans-Isthmian Canal

As Carryall faded from the excavation program's prospectus, the AEC's role in the Central American canal studies was elevated. Although the canal commission was in charge of the studies, under its direction the AEC was responsible for both developing the nuclear excavation techniques and evaluating the safety of the potential nuclear construction job, including bioenvironmental as well as meteorological studies. Despite the fact that a practical "multiple operations" demonstration like Carryall was still seen as a necessary step in the canal studies, the canal commission was compelled to proceed with its evaluation of the Trans-Isthmian canal—as it was commonly called—with or without Carryall, because its final report was due in Congress in 1968. And yet, given the obstacles experienced in the evaluation of proposed demonstration projects, how could the question of safety for a project as massive as the Trans-Isthmian canal—150 times larger than Carryall—be approached? And how was the technical problem of safety, as it was viewed by advocates of nuclear excavation, to be factored into the evaluation of the project's feasibility?

At a 1964 Plowshare symposium, medical physicist John Gofman, head of Livermore's new biomedical division, noted the difference between feasibility, in the technical sense, and the biological uncertainties associated with nuclear excavation. As Gofman understood it, "technological advances in the physical and engineering sciences have made a variety of projects feasible, while the biological and medical knowledge requisite to assess 'biological cost' is far from adequate."[56] Livermore's biomedical division had been established largely in response to the radioiodine controversy of the previous year. By 1965 Gofman had informed the Livermore leadership that, "My view is that building a canal with hydrogen bombs would be biological insanity." Gofman later recalled, "Edward Teller was beside himself. I then heard, just joking but only half joking, 'John Gofman is the enemy within.'" Less than two weeks after Gofman criticized Plowshare, the AEC attacked him as personally incompetent. Gofman nevertheless persisted in his oppositional studies of the effects of low-dose radiation from his position inside the Livermore laboratory, where he worked arduously to debunk notions of "thresholds" and "clean" bombs until 1970, culminating in a highly publicized resignation.[57] But while Gofman turned his investigations toward the long-term biological and genetic hazards of low-level radiation, the general direction of Livermore and AEC thought in this area was just the opposite. In order to evaluate the potential costs of nuclear excavation projects, Plowshare advocates sought to collapse the problem of biology into a discourse of "safety feasibility." Biology, in other words, was to be considered from an engineering standpoint.

Speaking at a 1966 Smithsonian Institution Conference on Tropical Biology in Panama City, Panama, the AEC's Richard Hamburger set forth a careful definition of what was meant by "safety feasibility" in the context of the Trans-Isthmian canal study:

> The term "feasibility"—as I am using it with respect to safety—includes the identification of those effects of the nuclear explosion which might be hazardous; the assessment of these effects to determine if they do present possible unacceptable hazards to man and his environment; ways to avoid or control those effects which might present possible hazards; and an estimation of the operational costs associated with these measures. The word "safety"—with respect to detonation of nuclear explosives—means the determination that the nuclear detonation can be accomplished without injury to people, either directly or indirectly, and without unacceptable damage to the ecological system and natural and man-made structures. . . .
>
> In determining safety feasibility, therefore, several things must be done. First, an evaluation must be made that the design of the project is such that the safety definition which I have given can be fulfilled. This requires consideration of such items as meteorology, geology, seismology, hydrology, structural response to ground motion, acoustic waves, radiation, and the biological and ecological systems.[58]

Safety feasibility did not mean that bioenvironmental hazards were to be ignored. Rather, it was the approach to the evaluation of hazards—the engineering approach—that was significant: hazards, and ecological damage, could be designated as acceptable or unacceptable; hazards were to be controlled; biological costs were to be determined. The discourse of safety feasibility was thus, above all, a discourse of control; its presumption was that the biological consequences of the unprecedented endeavor of setting off nuclear bombs to blast a sea-level canal connecting the Atlantic and Pacific oceans *could* be determined to an acceptable level of certainty.

Following an exchange of notes between U.S. ambassadors and the governments of Panama and Colombia in 1966, the canal commission was authorized

Table 5.2
Cost Estimates for Favored Sea-Level Trans-Isthmian Canal Routes by Nuclear and Conventional Excavation, 1966

Route	*Estimated costs: Nuclear excavation*	*Estimated costs: Conventional excavation*
Darien Gap, Panama	$750 million	$5.5 billion
Atrato-Truando, Colombia	$1.5 billion	$5.3 billion
Conversion of existing Panama Canal, Canal Zone	Not applicable	$2.2 billion

SOURCE: AEC, "Canal Study Commission Comments on Katzenbach's Questions," November 17, 1966. AEC 811/143. DOE doc. NV0075867, Nuclear Testing Archive, Las Vegas, Nevada.

to engage in on-site "safety feasibility surveys" in those countries. Before this fieldwork began, along with new geological and engineering surveys, the route through Panama's Darien Gap was still believed to be the most cost efficient if nuclear excavation was to be utilized, followed by the Atrato-Truando route in northern Colombia (see figure I.1, routes 17 and 25; see also table 5.2). If conventional methods were used, the transformation of the existing canal to a sea-level structure was considered the cheapest alternative. Because of the proximity of Panama City to the Canal Zone, nuclear excavation there was no longer considered an acceptable option. Full-scale field studies in Panama and Colombia were initiated in January 1967.[59]

COLUMBUS AND THE CUNA

Under the broad—but now, after Hamburger's speech, clearly defined—category of safety feasibility, AEC surveys for the Trans-Isthmian canal included everything from meteorological studies designed to predict fallout distribution to investigations intended to help predict ground motion and air-blast effects to analyses designed to assess the potential bioenvironmental effects of radiation. Many of these field studies were contracted out to other scientific agencies, to be coordinated by the AEC, which would, in turn, evaluate and integrate their work in order to ultimately provide the canal commission with a basis for making its final evaluation of the nuclear canal. The U.S. government–operated Environmental Science Service Administration was invited to carry out a two-year program of meteorological studies in Panama and Colombia, while the Battelle Memorial Institute (BMI) of Columbus, Ohio—an Ohio State University–based organization with much previous experience carrying out biological fieldwork for the AEC at the Nevada Test Site and Hanford Nuclear Reservation in Washington—was contracted for the bioenvironmental studies.

BMI directed studies, partly through subcontracts of its own, in six primary areas: (1) marine physiochemistry and biology; (2) fisheries resources of the estuaries and oceans; (3) terrestrial, freshwater, agricultural, and marine ecologies; (4) radiohydrology; (5) human ecology; and (6) radiation-dose estimation. Taken together, this information was to be used to "determine radiological-safety feasibility" of the proposed nuclear canal.[60] But how much information was enough to make this determination? In practice, the production of scientific information for evaluating safety feasibility was to be quite limited, with fieldwork restricted to eighteen to twenty months (and possibly as little as one year) of observation. This six-pronged approach was "generally agreed upon" by biologists at one of two BMI conferences held in Columbus in 1966, at least according to Alfred Klement, Jr., the AEC's new scientific liaison director for the canal studies. Some concerns, however, were expressed regarding the rather circumscribed extent of the biological program. Klement reported: "While most scientists advocate a larger program, most all of those contacted feel the approach sound and realistic from the standpoint of a feasibility study. Most scientists would advocate a much larger program to study all environmental effects if feasibility is shown and preparation for construction were approved. A number of groups having more basic research as goals would advocate beginning such studies during the feasibility phase."[61] If feasibility could be established with more limited studies, then scientists hoped that a more extensive program of pre- and postshot surveys could later be designed to facilitate a proper biological experiment.

But if Carryall had depended on numerous contingencies, from treaty considerations to a much accelerated rate of progress in clean-device development, then such contingencies increased by several orders of magnitude for the Trans-Isthmian canal project. Moreover, given the scale of the contemplated canal, it was evident that even if the canal commission found the project feasible, it would undoubtedly be subject to further review by agencies such as the National Academy of Sciences, the Federal Radiation Council, and perhaps the United Nations Scientific Committee on the Effects of Atomic Radiation.[62] Given the obstacles that beleaguered Plowshare already faced on many fronts, even establishing so-called safety feasibility may have seemed a long shot for all concerned. The biologists clamoring for a larger program, in this sense, may have hoped to maximize their research funding at the front end of the canal studies, while the AEC, in the matters of biology and ecology, preferred a cautious fiscal approach. Because the project faced so many obstacles, there was less directly at stake in the canal bioenvironmental studies than for those intended to precede Chariot, or perhaps even those to be organized to precede some future demonstration project. And yet as part of a wide-ranging, twenty-two-million-dollar canal commission study (on which the entire excavation program was now depending), the AEC's effort to secure safety feasibility for its geographical engineering project—the very exercise—was vital for Plowshare for its own sake.

Critical to the assessment of safety feasibility, as fieldwork in the tropics began in January 1967, was the development of a human ecology program to identify "potential pathways of human exposure to radioactivity," to predict external and internal radiation doses, and to "evaluate the potential need for, and probable effects of, protective actions that may be required to insure the radiological safety of human populations and/or to minimize their potential exposure to external and/or internal radiation." Field studies in human ecology, in accord with these goals, were designed to acquire data in four categories: (1) demographic characteristics of human populations in the region, their "cultural affinities," and geographical distribution; (2) dietary habits and subsistence activities of each "subpopulation"; (3) characteristics of material culture, occupations, and other activities that bring individuals into contact with their environment; and (4) "other attributes of each subpopulation's nature—social, political, economic, psychological, internal, or external migrations, etc.—which have a bearing on the feasibility and probable effects of protective actions such as evacuation and resettlement which may be required to insure their radiological safety or to minimize their potential exposure to radiation."[63] The human ecology studies were thus designed to predict not only the probable effects of radiation exposure on the "subpopulations" in the event of nuclear canal construction, but also to predict populations' social and political behavior in a manner intended to facilitate the AEC's safety program.

Human ecology studies in Panama were undertaken in five regions, as classified by BMI in terms of the racial and ethnic groups who lived there (see figure 5.3). These included two separate groups of Cuna Indians (areas A and B on the map), Choco Indians (area C), and two groups of "Negro and mestizo populations" (in areas D and E) who were composed primarily of frontier colonists who had moved from more populated areas of Panama and Colombia, respectively.[64] All of these groups were engaged in subsistence agriculture, while the colonists also farmed bananas and plantains and harvested timber commercially. The Cuna and Choco made their livelihoods exclusively from the land and waters of their environments: fishing, hunting, slash-and-burn agriculture, and gathering activities accounted for their primary food sources. These practices, of course, made them especially vulnerable to radioactivity as it moved through the food chain.

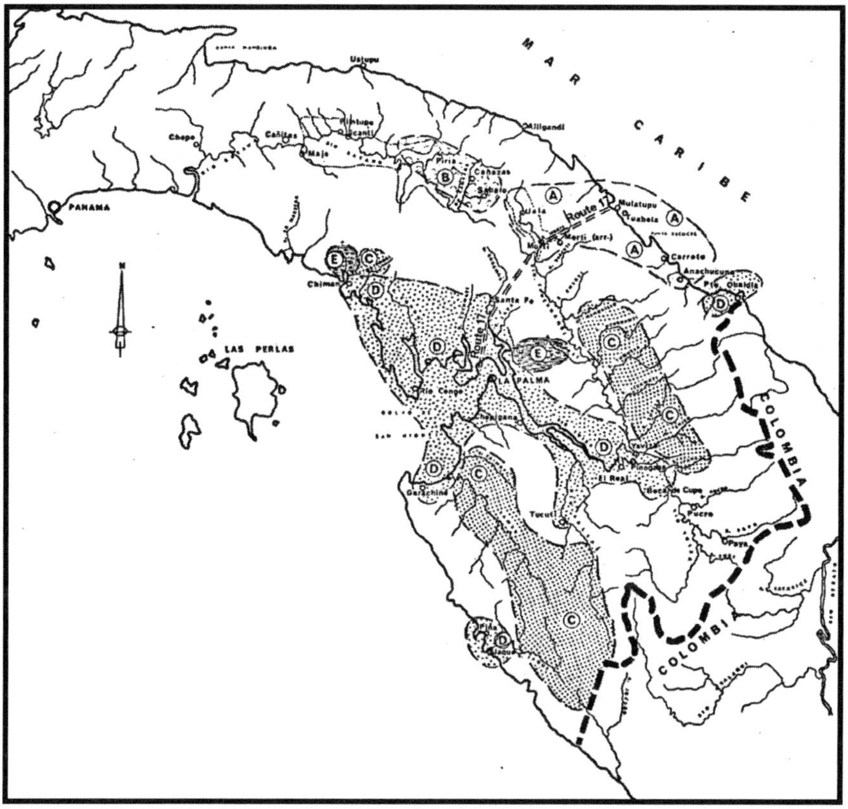

Figure 5.3. Proposed areas for human ecology studies, Darien, Panama. The Battelle Memorial Institute of Columbus, Ohio, was contracted by the AEC to direct ecological field studies as a part of the Central American nuclear canal studies. Regions A–E are identified by an associated population: (A) Cuna group from the Upper Chucunaque and San Blas; (B) Cuna group from the Upper Batano River; (C) Chocoe group; (D) Negro group; (E) Human groups originating from western Panama. Broken lines indicate existing and requested Cuna reservations; nuclear canal Route 17 is labeled in double broken lines across the center of the map. *Source: Battelle Memorial Institute (1967), "Bioenvironmental and Radiological-Safety Feasibility Studies, Atlantic-Pacific Interoceanic Canal," DOE doc. NV0016493, Nuclear Testing Archive, Las Vegas, Nevada.*

The Cuna living in villages on the upper Rio Chucunaque and its tributaries (in area A) were a special concern to the human ecologists, because of their location straddling the proposed canal route near the Caribbean side. As BMI scientists sought to carry out their field studies, however, they also expressed concerns about their own safety, for here, according to BMI's first report, the human ecology program's efforts to develop an integrative analysis of the social, political, economic, and psychological attributes would indeed be put to the test:

> The few visits that have been made to the region have demonstrated that these Indians are unfriendly to outsiders. Hostility has not extended as far as physical attack, but it is manifest in refusal to cooperate, in demands of payment for passage through or use of the area; and it may go as far as imprisonment of persons who have violated what they consider to be their "reservation." While they do not legally possess a reservation like that of the Cuna Indians of the upper Bayano and San Blas, they have made a topographic map of the region they occupy and proclaim it as theirs.[65]

In contrast to other Cuna groups encountered, this one was believed to be "politically independent and far more conservative and resistant to culture change." With some trepidation, it was added that "traditionally, they have had a reputation for belligerency."[66]

Despite these initial concerns, the canal commission reported to Congress in July 1967 that relations with "the Indian tribes in both Panama and Colombia have been generally good and continue to improve despite their initial hesitance to accept strangers." These good relations were facilitated by a liaison program that explained the data collection effort to tribal and village chiefs and also, the commission conceded, by more material negotiations and exchanges: many natives were employed as *portadores* and *macheteros* by the scientists along both routes, some medical assistance had been provided in the villages, "limited athletic equipment" was distributed, and local inhabitants had been encouraged to make use of new trails and boat landings constructed for the purposes of the safety surveys. "The Indians have consented to the removal of specimens and rock samples from their lands," it was noted, "and selected individuals have permitted blood samples to be taken for study purposes."[67] But just what else would they consent to?

Preliminary safety plans, set forth in the canal study commission's 1967 report to Congress, called for the evacuation of more than thirty thousand people from either the Darien or the Colombian route.[68] More than a "fallout sector," these plans constructed "exclusion zones" at least 40 miles wide by 100 miles long (see figure I.1). In turn, the plan called for the resettlement of these people and, presumably, some manner of provision for them while they waited, several years at least by any standard, for their homes to be deemed safe. Even leaving aside, for now, the AEC's problematic and contested definitions of safety—which made the assertion that eventually the evacuees could return to their traditional agricultural and hunting lifestyles without severe somatic and genetic cell damage rather questionable—and leaving aside the "normal" sorts of problems the AEC anticipated for a large-scale evacuation, such as dealing with the ill, the aged, and the very young (as described for the Tennessee-Tombigbee project), the canal evacuation plan would also have been likely to face more novel tactical problems. According to one critic, the Cuna could be expected to "resist attempts to evacuate them and undoubtedly many will hide in the impenetrable jungle nearby. Some who are not found and forcibly evacuated would be killed or injured by the nuclear explosions."[69]

Whatever value the medical assistance and limited sporting equipment provided, and however dubious BMI scientists believed was the Rio Chucunaque Cuna's map of sovereignty, it was clear that the Cuna would be difficult partners to enlist in the project of geographical engineering.

CRITICAL BIOLOGISTS

In September 1968, the American Institute of Biological Sciences and the Ecological Society of America sponsored a special symposium on the canal project at Columbus, where BMI and other contract scientists presented much of their research publicly for the first time. According to one science beat reporter, reflecting the tone inflected by a group of "critical biologists" at the meeting, BMI's safety feasibility study was "astonishing for its knowledge gaps."[70] Although new configurations for the layout of the explosives were described at the symposium—engineering and geological studies now indicated that 437 thermonuclear explosives with a yield of 275 megatons were needed to excavate the Darien route, and 223 explosives for 245 megatons for the northern Colombian route—much of the data concerning the amount of radioactivity expected to enter the food chain remained classified. While BMI project leaders acknowledged that, because of the uncertainties associated with the fallout problem, their evaluative role would be limited, problems in BMI's coordinating role were compounded by a lack of integration of the studies. Anthropologists, for example, presented research on the fishing and agricultural practices of native communities, but said little about the possible effects of radiation moving through the food chain. The issue of human uptake of radioactive elements *was* explored by health physicists from the Oak Ridge National Laboratories, who used computer codes to model the entire radio-ecological system. But these models were criticized at the symposium for basing physiological data on a 154-pound, white American "standard man," ignoring the fact that this physiology would likely be very different than that of the "standard man" for the Cuna and Choco, who averaged 120 pounds, let alone that of the women and the more vulnerable infants and children.[71]

The radiation safety studies had been further hampered by budget limitations. The Environmental Science Service Administration's weather data collection along the proposed routes was cut off at eighteen months and, with it, research programs in hydrology and medico-ecology. Along both routes, however, it was clear that heavy cloud cover and precipitation throughout much of the year would increase the fallout hazard; rainfall at the Darien weather stations ranged from 68 to 102 inches per eleven months (excluding data for the heavy rains of July), and from 268 to 433 inches annually along the Colombian route.[72]

BMI's William Martin, director of the bioenvironmental program, echoed many of the statements that the AEC's John Wolfe made six to eight years earlier about Project Chariot. Martin assured biologists at the Columbus meeting that even though the canal commission's feasibility report was due in Congress the following year, the studies were still quite preliminary, and that the blasts would only be approved if radiation safety could be ensured to the extent no one would be exposed to radiation levels in excess of internationally accepted standards.[73] But even if such assurances were made—by Martin, Oak Ridge modelers, or perhaps the Nuclear Cratering Group—whether or not other scientists would believe them was by this time another matter.

Experimentation, Trust, and Proof of Canal Feasibility

Despite earlier proclamations that nuclear excavation had become a matter of national policy, the politically beleaguered Johnson administration repeatedly postponed excavation tests. As one insider put it, Johnson did not wish to open up another "propaganda front."[74] But in 1968 three more nuclear excavation tests were carried out (see table 5.3). Cabriolet, in January, was another

Table 5.3
Nuclear-Cratering Experiments at the Nevada Test Site

Date	*Code name*	*Yield (kt)*	*Soil medium*	*Classification*
03/23/55	Teapot Ess	1.2	Alluvium	Military
10/14/58	Neptune	0.1	Tuff	Military
03/05/62	Danny Boy	0.4	Basalt	Military
07/06/62	Sedan	104.0	Alluvium	Plowshare
12/18/64	Sulky	0.1	Basalt	Plowshare
04/14/65	Palanquin	4.3	Rhyolite	Plowshare
01/26/68	Cabriolet	2.3	Rhyolite	Plowshare
03/12/68	Buggy	5.5 (5 × 1.1kt)	Basalt	Plowshare
12/08/68	Schooner	35.0	Rhyolite	Plowsare

low-yield, low-publicity test in Nevada. In March, a reduced version of the Buggy row-charge was completed, consisting of five 1-kiloton explosives set off in a line, a first run at the nuclear canal concept. The last of the three would be Schooner, originally planned as a 100-kiloton blast in granite in southern Idaho, a test that by this time had been reduced in size and shifted back to the Nevada Test Site, although the Nuclear Cratering Group continued to make use of the granite formation at the Idaho site for its high-explosives experiments. All three nuclear tests were supported under the canal excavation studies, part of an eighteen-million-dollar Plowshare budget (1968 dollars) for 1968 that represented the peak of Plowshare spending to date (see figure 4.4).[75] Although each of the experiments was announced publicly the day before detonation, all remained decidedly private events, restricted to authorized personnel, and definitively unspectacular. The "international observer program" that Chairman Seaborg had foreseen in his 1963 congressional testimony had never materialized.

Bold claims were advanced nevertheless. The headline of a May 1968 *Washington Post* story announced: "Buggy, Cabriolet A-Test Prove Canal Feasibility." Citing an unidentified AEC source, the article reported that a new "Ivory Soap" thermonuclear bomb, with a fission component of just .56 percent, had been employed with great success in the earthmoving tests. Fallout was contained within just a few hundred yards of the craters (in the case of Buggy, a 900-foot-long ditch), and "radiation right at the source could not even be detected three days after the shots."[76] The only technical impediment to the feasibility of the Trans-Isthmian canal, at least according to the *Post* reporter, was the problem of scale: the Ivory Soap design principles simply had to be transferred to larger explosives.

The authors of the July 1968 Fourth Annual Report of the canal commission were somewhat less sanguine about the future of the project. Although the report claimed that it was "highly probable" that the sea-level canal excavation would be deemed "technically feasible," it also recognized "the possibility of large contingency costs, particularly in nuclear excavation at the present level of the technology." While the question of whether or not project proponents had the right to move tens of thousands of people in Panama or Colombia from their homes was not raised (with political and emotional considerations duly excluded from the report's engineering standpoint), it was clear that it would cost a great deal to do so. Whatever breakthroughs were evident in clean-device development—progress of which the canal commission was apparently unaware—the report stressed that more experimentation in nuclear cratering was needed to properly evaluate the nuclear option. Guarding against future delays in the experimental program, the commission stressed that "timely execution of the remainder of the Atomic Energy Commission's planned program of nuclear excavation experiments is essential to develop more reliable cost estimates and to demonstrate the technical feasibility of nuclear excavation." To make up for delays already experienced, the commission asked for and received an extension for its final report to December 1, 1970, citing the need for more Plowshare experiments to support its study.[77]

But relatively small experiments like Buggy and Cabriolet could go only so far in proving canal feasibility, at least for the canal commission. Commission chairman Robert Anderson thus informed AEC chairman Glenn Seaborg that the remaining experiments in the excavation program would have to be completed in short order for a determination to be made about the nuclear canal, and he urged Seaborg to seek funding to execute no fewer than five experiments by the end of fiscal 1971.[78] As the proposed excavation tests listed in table 5.4 indicate, however flexible and resilient the experimental program had been up to this point, there was by this time little demand for anything but the large-scale nuclear excavation tests. In addition to Schooner, Projects Yawl, Galley, and Phaeton were all believed necessary to produce new data on row-charge explosions and cratering in hard rock at yields high enough that they could be extrapolated for the canal cuts, which would potentially employ explosives as large as 5, 10, 25, and 50 megatons.[79] Project Gondola, a cratering experiment to be conducted in wet clay (at a site as yet undetermined), would be particularly germane to soil conditions in Panama's Chucunaque valley, where field studies had left Army Corps engineers increasingly concerned about the problem of slope stability.

Seaborg, for his part, was well aware of the need for more experimental data, and he was outwardly optimistic that they could be produced. Despite Plowshare's "previous inability to obtain approval for . . . nuclear cratering experiments," he advised Anderson, the problem seemed "to have been alleviated for appropriately designed experiments."[80] The year 1968 was a busy one for Plowshare scientists in Livermore and Nevada. And perhaps even more significant, serious discussions for a harbor demonstration in Australia—conforming roughly to the proposed Galley experiment (see table 5.4)—were well under way. Whether or not it was realistic to believe that the entire program outlined in table 5.4 could be completed in a timely fashion, Plowshare pressed on in its quest for experimental funding and, in turn, in its efforts to prove the feasibility of the tropical canal excavation through experimental earthmoving blasts in the desert.

When John Kelly, coming before the JCAE in 1967, had requested funds for Plowshare for the following year, he claimed that total airborne fission products in cratering explosions "should be about 100–150 times less than the amount estimated to have been released from the 1962 Sedan experiment." But further

Table 5.4
Experimental Program in Nuclear Excavation, August 1968

Fiscal year	*Project code name*	*Description of experiment*
1969	Schooner	Forty-kiloton point charge (crater) in hard rock
1970	Yawl	Several-hundred-kiloton point charge in hard rock
1970	Galley	Row-charge experiment in varying terrain involving simultaneous detonation of about seven nuclear explosions in yields ranging from a few kilotons to a few hundred kilotons
1970	Gondola	Cratering test to be conducted in wet clay shale
1971	Phaeton	One-megaton point charge in hard rock

SOURCE: Anderson to Seaborg, August 23, 1968. Included in DOE doc. NV0173161, NTA.

experimentation was essential to continued progress, Kelly told the JCAE, and to push nuclear excavation to the brink of practical applicability. "Verification of this prediction . . . in the yield range contemplated for the Schooner experiment," he continued, "would be a dramatic demonstration that radioactivity should pose no real health obstacle to large scale nuclear excavation."[81]

On December 8, 1968, after waiting two days for "favorable winds," the now 35-kiloton Schooner explosive, emplaced 350 feet underground in a shaft drilled into the hard rock of Area 20, was detonated at last. The blast slightly exceeded predictions for quantities of earth moved, producing a crater 200 feet deep and 850 feet wide. A low-key public announcement followed two days later, indicating briskly that Schooner would "provide information about cratering effects in hard rock."[82] But of course there had never been much doubt that Schooner would produce such new information, or, for that matter, that nuclear explosions could move tremendous volumes of earth and rock. AEC reports subsequently added that the amount of radioactivity released from Schooner was "infinitesimal."[83]

Not so, according to Robert Pendleton, the University of Utah radiological health scientist who had, six and a half years earlier, observed the Sedan dust cloud quite by accident in the Wasatch Mountains and then taken it upon himself to publicize radioiodine risks to Utahans. Pendleton notified the press that Schooner had in fact produced the highest levels of fallout recorded in Utah since 1962. Although Pendleton noted that, unlike 1962, the quantity of fallout was not great enough to suggest a major contamination of the state's milk supply, he insisted that the fallout would compound existing hazards by raising radioactivity levels, elevating risks for lung cancers, lymph node carcinomas, and other malignancies. "The people of Utah are once more being showered with radioactive materials," Pendleton charged "a very serious type of pollution which will remain with us for years to come." AEC reports describing Schooner fallout as harmless were, as Pendleton saw it, "gobbledygook . . . the sort of thing they have been telling us since 1962." Indeed, for Pendleton, there was, by this time, a "credibility gap between the AEC and the knowledgeable people of the western United States."[84] The AEC responded to Pendleton's comments the next day in a new press release provided to the Las Vegas AP and UPI bureaus as well as two Salt Lake City newspapers, maintaining that a "special minimum-fission nuclear explosive" had been used for Schooner, and that as a result, no radioiodine was detected in milk supplies. It was added that, according to John McBride, deputy director of the Public Health Service, not only did present levels of radioactivity not constitute a hazard, but further, "any statement implying such a hazard should be considered as 'irresponsible.'"[85]

Who were the knowledgeable people of the West to trust? Billionaire Howard Hughes, for one, an increasingly hostile neighbor to the Nevada Test Site, wanted to hear more from Dr. Pendleton, and Hughes's public offer to fund Pendleton's research kept the story alive in the media.[86] Mrs. Lucille Bushnell of Provo, Utah, was equally concerned, and she telephoned the AEC's Office of Public Affairs (at the urging of one of Pendleton's assistants) to question the discrepancy over Schooner fallout. Bushnell was assured that the levels of radiation around Provo "did not present a danger," and it was noted for the AEC files that "Mrs. Bushnell seemed satisfied and somewhat surprised at our candid and apparently honest replies to her concern."[87]

Such candor had its limits when it came to the public disclosure of data from Schooner. In reviewing a preliminary draft of the AEC's annual report to Congress, Seaborg sent it back, indicating that there was "too much detail on radiation off-site for NTS." Paragraphs on the "highest levels" were to be removed, and instead Schooner was simply to be added to a list of tests in which radioactivity had been "detected off-site."[88] Not mentioned was that, as a direct result of Schooner, radiation increases in Canada signified the most egregious U.S. violation of the LTBT to date. Although radiation at several Canadian sampling stations rose from ten to twenty times normal background levels from December 13 to 15, as reported the following month by the Canadian Broadcast Company, neither the AEC nor the Canadian government commented on the boundary violation. *Pravda* reported the news, but the Soviet Union, which had itself recently vented radiation over Japan, did not challenge the United States over the matter.[89] Data regarding exactly how much fallout was released—the verification of predictions that Kelly had called for in front of Congress—were not discussed until more than a year later, and the results fell far short of Kelly's optimistic predictions.[90] Even so, Schooner could be or, rather, had to be represented as yet another important stage of Plowshare's progress-through-experimentation. And in turn, this progress was incorporated into the discourse of feasibility and practicality that accompanied the Plowshare excavation program as it made its next—and ultimately, its final—geographical move.

Australian Proving Grounds

In December 1968, the premier of the state of Western Australia notified the national government of his state's interest in a proposal to examine the feasibility of using nuclear explosives to excavate a harbor at Cape Keraudren, on Australia's northwest coast. Western Australia was under agreement with the Sentinel Mining Company, an Australian subsidiary of the National Bulk Carriers shipping firm, to construct a port for the export of iron ore from the nearby Nimingarra deposits in the Hammersley Range, and Sentinel, the premier indicated, had expressed interest in using nuclear techniques for the harbor.[91] In turn, Australia's minister for national development, David Fairbairn, on January 23, 1969, requested the participation of the (U.S.) AEC and its Plowshare program in the feasibility study. As Fairbairn explained it, the new harbor facilities were needed to facilitate mineral development in the region, which lacked natural deep-water harbors, and "from the nuclear viewpoint" the Cape Keraudren site was ideal because its location on the coast of the Great Sandy Desert was singularly remote from population. "The proposal is being treated as urgent," Fairbairn added, because, in the first place, Sentinel had a schedule to meet for the shipment of the ore, and second, "because the spread of population is reducing the number of sites at which construction by nuclear blast can be carried out without interference with local activities."[92]

The nuclear design called for five 200-kiloton explosives to be placed in a row 1,100 feet apart and buried roughly 800 feet deep in the sea floor, and detonated simultaneously to create a long channel deep and wide enough for the largest ore carriers to navigate within. Unlike a conventional dredging process, the nuclear row-charge would also, in theory, produce walls of fractured rock along the sides of the channel, thus providing refuge from Indian Ocean typhoons. Of course, before any of this geographical engineering could take place, feasibility studies would be necessary to evaluate technical, economic, and safety aspects of the proposed harbor, and Fairbairn announced that, if an agreement were reached, the Australian Atomic Energy Commission (AAEC) would "work closely with American experts" from the Plowshare program. While it was acknowledged that underwater nuclear earthmoving had not yet been tested, and that Plowshare scientists had not yet utilized peaceful nuclear explosives outside U.S. territory, Fairbairn averred that Plowshare techniques "so far have proved markedly successful in other fields."[93] An initial feasibility study could be completed within a few months, it was announced, and nuclear blasting might commence in as little as one year's time.

Two weeks later, shortly after President Richard M. Nixon's inauguration, the United States surprised nobody by agreeing to participate in the feasibility study. Australian newspapers quickly reacted with enthusiasm over the nuclear harbor's heroic implications and with trust in the AEC's contention that radiation was "no longer a problem."[94] And it had all been set in motion, apparently, by the Sentinel Mining Company.

CAPE KERAUDREN PROJECT

Yet plans for the instant-harbor project for Western Australia had been initiated well before Fairbairn and Sentinel had signed on to them. American and Australian nuclear scientists had discussed possible geographical engineering projects for Australia since at least 1962, and plans for what became the Cape Keraudren nuclear harbor feasibility study were pushed ahead a few years later at least partly as a result of Edward Teller's networking strategies. In August 1966, Teller paid an "unofficial visit to Australia," as the AAEC described it, and in "the course of a privately sponsored tour" he visited several remote locations to assess their suitability for nuclear construction projects. In addition, Teller "had discussions with members of the [AAEC] Commission and various groups concerned with large developmental projects in mining, water conservation and other fields."[95]

Shortly after his return, Teller suggested a more detailed reconnaissance of harbor sites to the AAEC secretary, adding that "it might be an excellent idea if two of our experts from Livermore would be invited for a short (and unpublicized) visit to make a thorough investigation of possible sites. I believe that two weeks spent in the neighborhood of the Hammersley [*sic*] range should serve the purposes." A new friend of Teller's, Perth mining magnate Lang Hancock, who had acted as a private guide for Teller in his tour of the region, then contacted Livermore about his need for a harbor along the remote northwestern coast of Western Australia. And so Hancock visited with Teller and physicist Milo Nordyke at Livermore in August 1967 to discuss possibilities for the port facilities that Hancock was obligated to build in connection with a mineral rights agreement he had made with the Australian government. Hancock indicated that he hoped to have a new harbor in place within two years. As Nordyke countered, the "construction of such a harbor in the time frame of 1970 was not out of the question, but rather very desirable from our standpoint."[96]

Nordyke outlined for Hancock the appropriate steps to be taken to advance the project: first, Hancock should approach the Australian AEC, so that any initiative would be generated by the Australian government, and second, Hancock and Teller would make use of contacts to "develop a receptive attitude in the U.S. for such a project should the Australian government request it." Hancock was apparently quite concerned about the question of costs, but here Teller proposed an elegant solution: as Nordyke recalled, "Dr. Teller raised the subject of doing one of our excavation experiments at the site, thereby helping to defray some of the safety evaluation costs, which would be assigned to the demonstration project. This, of course, was very attractive to Mr. Hancock."[97] Unlike the recently abandoned Project Carryall, which required several large-scale experiments to precede it, the Western Australian harbor was thus seen from the start as an experimental exercise. Framing the harbor as both an experiment and a useful demonstration project, Teller and his Livermore colleagues, as with Chariot a decade before, sought both to try and to show their experiment at once.

In September 1968 Sentinel's Malcolm Reed visited Livermore to discuss the Cape Keraudren harbor and, more specifically, what the nature of his company's involvement might be. By October 14, after conferring with parent company National Bulk Carriers (which insisted rather vaguely that its costs be "limited"), it was decided to push the project forward through the Western Australian state government. The project was to achieve an international agreement for a feasibility study by March 1969, and postnuclear, conventional construction of the harbor was to begin by March 1970.[98]

As the physicist Nordyke saw it, this timing was perfect. Although the abbreviated planning and research phase would put the project on a tight schedule, Nordyke argued that the expeditious conduct of the project was precisely what Plowshare needed to prove itself in the "pragmatic engineering world":

> Through the years, we have found that timing is probably the most significant factor from the standpoint of acceptance or rejection of a nuclear excavation project by a commercial or government sponsor. If the excavation project is truly to be an integral part of an engineering project which will be in common use, it must, of necessity, be competitive with a conventional solution with respect to both time and money. . . . Thus, for any viable demonstration project, we must be prepared to execute it in a much shorter length of time than is the general practice in our experimental program, and to accept the technical limitations such an approach imposes.[99]

It was not just Sentinel's need to ship its ore that hurried Plowshare; Livermore and the AEC were running out of time to provide experimental data to the canal commission. The row-charge at Cape Keraudren, however rushed its execution might be, could produce four kinds of new data relevant to the Trans-Isthmian canal. Its aims, as Nordyke explained them, were:

1. To extend knowledge of the short- and long-term effects of water on row cratering
2. To investigate slope stability factors "under real conditions"
3. To extend knowledge of the spacing of explosives for row cratering at high-explosive yields
4. To provide data on the conversion of nuclear row craters into a practical engineering structure

The Cape Keraudren project was thus an experiment in every sense. But the spectacular aspects of the harbor demonstration were not lost on Nordyke either. "As important as these technical data would be to the nuclear excavation program," he added, "the greatest impact would almost certainly be in demonstration of the practicality of the nuclear approach, and the acceptance of nuclear excavation as a viable engineering tool by the governments of the United States, Panama, Colombia, and their peoples."[100] But what would the failure, once again, to demonstrate anything at all prove?

FEASIBILITY

With the help of an impassioned letter to President Johnson from Congressman and JCAE member Craig Hosmer (R-California)—in which Hosmer likened the Australian nuclear harbor to the Apollo 8 mission—$267,000 had been appropriated for the joint U.S.-Australian feasibility study well before Minister Fairbairn's request for AEC assistance.[101] With many of the early "paper studies" of the Western Australian harbor project already under way at Livermore, planning moved ahead rapidly. By February 1969, Livermore had produced a preliminary concept paper for the harbor scheme that included descriptions of several area sites, configurations of explosives, predicted effects of fallout, ground shock, air blast, and water waves, and operational planning procedures.[102] Meetings were held at the AEC's Nevada Operations Office on February 14–15, with representatives from the AEC, Livermore, the Australian AEC, and Sentinel assembled to discuss how the feasibility study was to be organized and executed.[103]

Of all the participants in the Nevada meetings, only Sentinel's Peter Stork conveyed any misgivings about the project. Expressing concern over the "high level attention" the project was receiving, Stork complained that his company was beginning to feel "trapped," as a result of Livermore's promotional efforts, into a commitment to the harbor project itself (rather than only to the feasibility study). But the others, apparently unwilling to let Stork's cold feet stand in the way of a nuclear harbor at Cape Keraudren, worked out an understanding that distributed responsibilities for the studies: nuclear excavation feasibility was delegated to the AEC; safety feasibility to both the AEC and the AAEC; and the areas of economic feasibility and harbor facilities construction to Sentinel. All studies would be integrated through an Advisory Committee, to be chaired by AAEC scientist Alan Wilson. Although it was acknowledged that the feasibility study might take longer than the three months allotted for it in early planning, September 1 (1969) was deemed a reasonable conclusion date, with detonation to follow nine to twelve months later.[104]

The primary safety concerns for the project, according to Wilson, had to do with groundwater and marine ecology, an ecology that included commercial shrimping in the area and the existence of nomadic Aborigines in the likely downwind area, although the precise nature of fallout sectors and exclusion zones for the project appears not to have been closely examined at the meeting. Nor were much larger population centers potentially downwind from the blast, like the island of Timor, considered in the February concept paper. Preliminary data suggested that winds tended to blow either offshore from Australia's northwest coast most of the time—toward Timor and Indonesia, and thus both a potentially severe health hazard and a violation of the LTBT—or southwesterly, toward ranching economies near Port Hedland, Western Australia, but Plowshare scientists remained as confident as ever that fallout hazards could be controlled.

For Livermore's H.C. Rodean, the very notion of a "feasibility study" was perhaps inappropriate to the task at hand.

> There is no doubt whatsoever that NVOO, LRL and USAEC contractors could go to Cape Keraudren and blow a big hole in the ground. This, in itself, is therefore not a question of feasibility. The real questions involve the value of such an operation to the U.S. as a demonstration of Plowshare technology and as a Plowshare research experiment, to the Australian Commonwealth and the State of Western Australia as a means to encourage regional development and as an acceptable civil construction method with respect to public health and safety, and to NBC/Sentinel/Premier as a timely and economical means of constructing a useful harbor. . . . Therefore, I regard the term "feasibility" report as unfortunate, and instead think of the AHFS report as a formal proposal to the top management of NBC/Sentinel/Premier and to the upper levels of the U.S. and Australian Governments.[105]

Rodean was absolutely right, from Livermore's perspective. The feasibility study *was* to be a proposal to do a nuclear construction job, and in the course of the study, its resourceful promoters would find a way to define it as safe, practical, and economical, perhaps even legal by some interpretation of the LTBT.[106] But while Rodean's position may have made good sense to scientists and engineers assigned or self-selected to participate in the Cape Keraudren studies, for others seeking to push the "proposal" through the U.S. and Australian governments, the rhetorical value of terms like *feasibility*, *demonstration*, and *experiment* remained important. And yet, just as had accompanied efforts to enact Chariot ten years earlier, whether the Cape Keraudren project was to constitute a demonstration or an experiment depended largely on the audience to which it was pitched.

Shortly after the February meetings in Nevada, Chairman Seaborg appeared before the Senate Foreign Relations Committee. The subject of the hearings was the ratification of the nuclear nonproliferation treaty (NPT), which contained a provision for nuclear states to provide peaceful nuclear explosive "services" to non-nuclear states, so that the latter could share in the purported benefits of the likes of Plowshare without developing nuclear weapons programs of their own. This provision was seen as a potential boon for the Plowshare program to facilitate its movement overseas. But as the question was put to Seaborg at the hearings (in relation to the Cape Keraudren project), was it at all problematic that Australia had not yet signed the NPT? No, countered Seaborg, because the Cape Keraudren project was an *experiment*, whereas the NPT provision of exclusive rights for PNE services among signatory states concerned already-working technologies. In early March, however, the American ambassador to Australia was compelled to assure Australians that there was actually "nothing experimental" about the project; everything proposed for Cape Keraudren had already been tested in prior Plowshare experiments in Nevada. Such contradictions and shades of meaning—between proposal and feasibility study, experiment and demonstration—would not go unnoticed as debates over the Cape Keraudren project gathered momentum in the public sphere.

ANOTHER FAILURE

Problems quickly emerged both inside and outside the experimental space delineated at the Nevada meetings. The involvement of the mining company Sentinel was essential to the pragmatic success of the feasibility study: neither the data concerning the conversion of the channel cut into a harbor nor the demonstrative purposes of the Cape Keraudren project as a practical experiment were possible without it. But Sentinel began to drag its feet. On the advice of John Kelly, Australian AEC secretary Maurice Timbs contacted Sentinel's Malcolm Reed on February 21 to find out the company's latest position on the project. Reed informed him that Sentinel—which had earlier insisted on the briefest of construction timetables—now needed more time for the assessment of ore deposits in the region; recent sampling indicated, Reed told Timbs, that some had been of lower grade than was previously understood. Sentinel required more information before it could proceed with the Cape Keraudren project. Timbs was taken aback by Sentinel's apparent reversal. As he recalled his response to Reed :

> I said I found this attitude extremely difficult to understand. The Company had done a great deal of work before it initiated the feasibility study by writing to the Premier of Western Australia. Nothing would have been done at the Government level unless the Company had taken this action. . . . [A]ction had [since] been taken to publicly commit three governments to the project and I said that it must be assumed that his action was not taken irresponsibly or as window dressing.

Reed saw it differently:

> He [Reed] said that the proposal for the feasibility study was the result of 2–3 years promotion by the U.S.A.E.C. Plowshare Division. He felt that it was clear that the objectives of the Plowshare Division were to get some Company to do something in the Plowshare field outside the United States and to say something publicly. Personnel from Livermore Laboratories had represented to Sentinel Mining Company that Niminjarra [the mountain range in which the ore was located] was a suitable project for Plowshare operations and it was they who proposed that Sentinel Mining Company study the feasibility of using a nuclear blast to make the harbor.

The problem, Reed claimed, was that the Australian government now appeared committed to the project regardless of the economics. Further, Sentinel (and National Bulk Carriers) "did not like all of the publicity associated with the project as this seemed to him [Reed] to imply a commitment to undertake the whole project."[107]

This publicity was soon amplified as scientists and politicians, from both inside and outside Australia, began to look more critically at the instant-harbor scheme. Even as Sentinel was beginning to hedge on its demands for the project's rapid execution, critics of Plowshare called attention to the proposed timing of the study—its apparent concern for construction timetables over ecological ones—as well as the unpredictable nature of the sea-floor blasts. Contra Minister Fairbairn's assurances to Australians of Plowshare's marked successes in numerous tests similar to the Cape Keraudren project, Marvin Kalkstein, a professor of earth sciences at the State University of New York at Stony

Brook, examined the program's experimental history and found no proof of safety. Instead, in a letter to the editor of the *New York Times*, he stressed the Western Australian nuclear harbor's many uncertainties: "The project is questionable on several grounds. There has not been sufficient testing of large nuclear explosives for earth moving to warrant proceeding with an application; and what testing there has been has indicated difficulties in containing the radioactive debris. Results of cratering tests in Nevada alluvium to date might also not be directly applicable to a deepening operation beneath the ocean floor in Australia." Kalkstein added that if the radioactive venting from Project Schooner was any indication, then the Cape Keraudren project was likely to break the LTBT by distributing fallout across Indonesia or other archipelagos. But Kalkstein was most concerned by the urgency of the project and the insufficient time allotted for establishing technical feasibility, a schedule that seemed, to Kalkstein, to be heavily influenced by Sentinel's demand for port facilities: "Thus, we have here a case where the commercial plans of a private enterprise appear to have taken precedence over treaty obligations and present limitations of the technology."[108]

Whether it was Sentinel's timetables or Plowshare's that were driving the Cape Keraudren project, and regardless of how long it would take Plowshare scientists to assert its feasibility, Kalkstein's larger point—that the project must be approached with caution, not urgency—resonated with some in Australia. On February 25, Kalkstein's letter was quoted by a member of the opposition Labor Party in the Australian Senate, and the same day, a West Australian Labor Party MP questioned the accuracy of fallout predictions for Cape Keraudren in the Australian House of Representatives. One month later, Opposition Leader Gough Whitlman held a press conference in Sydney to raise further doubts about the timing of the feasibility study by quoting a 1964 Australian AEC report which stated that ecological studies for nuclear excavation projects should cover at least a one-year seasonal cycle.[109]

Just days later, the AEC and its Australian counterpart put the controversy at bay when they announced in a joint press release that the feasibility study had been called off, for the time being:

> The Sentinel Mining Company has just completed a reevaluation of its opportunities in the mining and marketing of iron ore, which was to be the principal product shipped through the proposed harbor, and therefore desires to limit its participation in the proposed study.
>
> Since the economics of the mining venture is an essential element there is insufficient basis for proceeding with this study.
>
> However, the United States and Australian Atomic Energy Commissions continue to be interested in the possible use of nuclear ex-

> plosives for harbor construction and will, therefore, continue their review of practicability of applying this technology to other possible harbor sites in the area.[110]

Whatever the status of its need for a harbor in Western Australia, Sentinel plainly did not care for the publicity of the endeavor. But nor, for that matter, would the AEC have relished an international controversy centered on the risks and uncertainties of its nuclear excavation technology.

Even as the Cape Keraudren project was imploding at meetings in Washington, D.C., in late March 1969, a new challenge to geographical engineering in Australia was emerging in Sydney. Sydney University biologists L. C. Birch and N. A. Walker and physical chemist A. E. Alexander began to circulate a petition to five hundred Australian scientists attacking the Commonwealth's handling of the Cape Keraudren project.[111] In their efforts to learn as much as possible about the uncertainties and risks of nuclear excavation, Birch and Walker had sought out the 1961 special issue of *Nuclear Information* evaluating the risks of Chariot, and found that the only copy available in Sydney was located at the Australian AEC's library. Yet when the biologists requested the issue, they were told that it had "disappeared." Nevertheless, with a copy of the special issue eventually in hand (courtesy of the editors of the journal, by this time renamed *Environment*), the Sydney scientists constructed a petition of six demands, six *rights to knowledge* meant to ensure a more cautious—and more open and public—approach in the future, not just for Cape Keraudren, already defunct, but for any future "nuclear explosion projects" in Australia. Signed by 259 scientists in Australia, the list of rights was published in *The Australian* in May 1969:

> 1. To provide at the outset for a full ecological study of any land and sea areas to be affected, for a period of at least two years, before decisions to proceed with a nuclear explosion project.
> 2. To publish the results of this study in full.
> 3. To disclose the inventory of radioactive isotopes that would be released and their amounts, and the accuracy of these estimates.
> 4. To disclose the expected physical hazards and the accuracy of their estimation.
> 5. To appoint an independent committee to decide on the safety aspects of such a project when the above information is available, and when the risk posed by each isotope to be released can be reasonably estimated.
> 6. To constitute this committee of experts in relevant fields—from physics to anthropology—who are not in the employ of the government or government agencies, the United States AEC or interested commercial enterprises.[112]

Resonant with Don Foote's angry letter of protest to the AEC's Committee on Environmental Studies for Project Chariot written almost a decade earlier, the Sydney scientists insisted that, given the scientific uncertainties involved, sustained ecological observation must be matched by an institutionalized division of powers in the evaluation of the data and, accordingly, by accessible information and open public debate.

For a time, discussion continued among the Cape Keraudren project builders of transforming the harbor project—for which a feasibility study had still been approved by two governments—into a joint "general feasibility study" of several possible harbors in northwestern Australia, a move which, as the Australian AEC's Timbs saw it, would "keep up the momentum created by the Cape Keraudren proposal."[113] To these ends, two Australian journalists were given a tour of the Nevada Test Site in July 1969 to see the craters and to promote a more receptive attitude in Australia.[114] But in the climate of critical scrutiny engendered by Sydney biologists and Labor MPs, a new proving grounds was never opened in Australia, the geography most suited, it had seemed, to geographical engineering. Ultimately, if the Australian feasibility study had shown anything, it was that the *kind* of feasibility engendered by the AEC technocracy could itself be opened to challenge on a public and scientific basis.

For Plowshare, desperate for large-scale experimental results and, no less, a feat of actual geographical engineering to show off to the canal commission, the failure to demonstrate was a failure, period. It was supposed to have been a breakthrough year. Following the three cratering experiments conducted at NTS during 1968, and with the promising Australian project seemingly under way, international conferences had been planned for Tucson ("Education for the Peaceful Uses of Nuclear Explosives") and Las Vegas ("Public Health Aspects of Peaceful Nuclear Explosives") in March and April 1969. These conferences were designed to replicate the scientific and technical knowledge of the program, whether among public health contractors or through university education, and to facilitate the movement of Plowshare projects from the controlled spaces of Livermore and Nevada to those controversial places where the engineering standpoint of the program was viewed more critically. But while Edward Teller's comments in Tucson, presented at the beginning of this chapter, may have been taken to heart by the engineering educators, who were rapidly ushering nuclear excavation components into their nuclear engineering courses, the discourse of feasibility for nuclear excavation was increasingly discredited outside the community of contract laboratories and engineering firms, engineering departments, atomic energy commissions, and other nuclear cratering groups. Call it a lack of trust.

Chapter 6 Epitaph

Technocracy, Geography, and the Rights to Knowledge

Despite the canal commission's need for Plowshare to complete its experimental program in order to seriously consider the nuclear canal option, no new nuclear excavation experiments were funded by Congress after Cape Keraudren. As Edward Teller stepped up to the microphone at a Houston press conference in September 1969, progress in the nuclear earthmoving program seemed at last to have reached a point of diminishing returns. Amid the excitement of Alaska's North Slope oil boom, however, Teller announced that plans were under development at Livermore for the use of nuclear explosives to excavate an "atoll-style" harbor offshore from Prudhoe Bay to serve oil tankers. But Teller's perennial nuclear optimism was tempered this time by a dose of political economic realism. Funding for the project would have to come from the oil companies, Teller said, because "the Bureau of the Budget would never give us the money, so private enterprise will have to do it."[1]

By this time, Alaska was an oil state; after 1967, oil surpassed fisheries as the leading source of state income, and the September 1969 auction of 450,000 acres adjacent to Prudhoe Bay had netted the state an unprecedented nine hundred million dollars.[2] The necessity of getting this oil to market, for Humble Oil (Standard), Atlantic Richfield, and British Petroleum, meant exploring an array of transportation solutions, among them Livermore's latest instant-harbor scheme for the Arctic. In a case of geographical engineering emulating nature itself, this plan, which called for a single 200-kiloton explosion under the sea floor some twenty miles offshore, was designed to produce crater walls around a new harbor that would resemble one of the Pacific Marshall Islands atolls, where the Manhattan Project's and AEC's first postwar proving grounds had been established. The atoll, Livermore physicists conjectured, would provide some protection for oil tankers as they loaded their cargo under Arctic conditions.[3] But this nuclear solution, which relied on the concomitant development of new ice-breaking commercial oil tankers, was soon cast adrift. On the northern fringe of U.S. territory, in a region already parceled out to friendly corporations, nuclear excavation turned out to be a rather marginal venture. More interested in getting their oil to market than in demonstrating the viability of nuclear excavation techniques, the oil companies threw their weight behind the Trans-Alaska Pipeline to Valdez instead, and the second Alaskan instant-harbor concept remained, like Chariot before it, only an imagined geography.

With the collapse of this project, in the wake of the Western Australian debacle, the faded promise of nuclear excavation must have been clear even to Teller himself. Perhaps if, as he had suggested at the 1959 Plowshare Symposium, nuclear tests could be carried out on the moon, then experimentation in nuclear excavation might have continued. But by the fall of 1969, even after Neil Armstrong's historic moonwalk made such a lunar experimental space somewhat more feasible, there was no space left for nuclear excavation, not even in the marginal geographical spaces of what the Livermore physicist Milo Nordyke had called the "pragmatic engineering world."[4]

In January 1970 the Army Corps of Engineers issued a status report which, though it deferred any conclusions or recommendation to its final report (due in December), clarified some of the major problems that the canal commission faced in assessing the feasibility of the nuclear canal and, as such, signaled the onset of the earthmoving program's final demise. From the tone of the report, the corps had been let down in its nuclear canal study—first by the slow pace of the experimental program, and second by the geology of Panama's Darien. Exploratory drilling in the Chucanaque Valley had revealed an extensive formation of wet clay shales that, because of potential slope stability problems, no longer appeared suitable for nuclear excavation methods. The route through the Darien Gap had previously been considered the cheapest of the nuclear alternatives, but now, partly because the route through the Chucanaque Valley would have to be excavated by conventional means, the cost estimate ballooned from less than $1 billion to $2.9 billion. The new plans for the Darien route still called for 250 nuclear explosives to be set off in a series of twenty-seven separate detonations, with some forty thousand people evacuated from the "exclusion zone" to meet preliminary safety feasibility standards. Even if "the economic burden imposed by moving them is not large—about $140 million," as Brigadier General R. H. Groves of the corps saw it, "the social and political consequences for Panama might make it unacceptable."[5]

Meanwhile, priced at $1.9 billion—only slightly less than the Canal Company's earlier estimates for the conversion of the present Panama Canal to a sea-level canal by conventional means—the route through northern Colombia was now considered the more economical alternative for nuclear excavation. Less was known about the geology and physical and human geography of the corridor, but it was evident to Groves that the "imposed burden" of evacuating roughly twenty thousand people from the exclusion zone there might contribute to many of the same "social and political consequences" as the Darien route.[6]

The canal commission report was careful to stress that nuclear excavation might still be the best method for making the new canal. And yet, as Groves noted, the "inability to conduct the tests needed to prove out nuclear excavation theory leaves large gaps in our knowledge, making it impossible for us to state unequivocally that nuclear excavating techniques are feasible." Groves's statement thus provided an early indication that the five-year, twenty-two-million-dollar canal study would not even fulfill its primary purpose—evaluation of feasibility—because of a lack of data. Still, since Groves believed that it was "likely that nuclear excavation techniques will be widely used in the future," he lobbied for completion of the experimental program designed in support of the canal studies. Even if a thorough review of the nuclear method was not possible before the canal commission's December 1, 1970, deadline, Groves did not want for the study to be seen as a failure. "We have made real contributions to man's knowledge," he insisted. "We must not permit them to be forgotten; we must continue to build upon them and to expend [*sic*] our knowledge."[7]

The AEC made one last push for the nuclear earthmoving program. In its fiscal 1970 budget request, the AEC begged twenty-nine million dollars for Plowshare from the Bureau of the Budget, and received half that amount (see figure 4.4). Optimistic to the end, the AEC asked for forty-four million in 1971, but received only eight million, none of which was allotted for nuclear excavation; the use of nuclear explosives for underground natural gas stimulation and oil shale recovery, in cooperation with industry, were by this time the only Plowshare applications for which an active experimental program was to be funded.[8] In March 1970, Livermore director Michael May went hat in hand before the JCAE to request that an additional ten million be added to the budget, half of which would be slated for high-yield excavation device development.[9] The denial of May's plea, in effect, ended Livermore's involvement with geographical engineering, not with a bang but a whimper. In July 1970 the AEC formally notified the canal commission that it would not complete the necessary tests to support the canal studies, and in December the canal commission, to the certain disappointment of nuclear cratering groups around the country, could only recommend that the sea-level Central American canal be constructed by conventional means.[10]

In the early 1970s, the idea of nuclear excavation was not dead at Livermore, and by no means was it ruled out, but nor was it funded through an active experimental program. Geographical engineering schemes, developed by university scientists as well as by private engineering firms, still graced the desktops and files of Livermore scientists, from a nuclear harbor in Malaysia to the transcontinental North American Water and Power Alliance (NAWAPA) project to an inland sea atop Egypt's Qattara Depression, designed to generate power as Mediterranean waters rushed in to fill the new sea. Perhaps Teller's presence on the board of Thailand's Kra Canal Commission raised the hopes of diehard Plowshare enthusiasts, but no evidence suggests that this or any other nuclear excavation project, by this time a liability, not a boon, to public relations, was seriously pursued under Plowshare after 1970.

Rehearsed in the politics of resilience, Plowshare had already begun to refocus its efforts underground in the areas of natural gas and oil shale exploitation. Underground gas stimulation experiments were carried out in New Mexico, in 1967, and near Rifle and Rio Blanco, Colorado, in 1969 and 1973, respectively. The Colorado experiments were designed as both commercial and technological demonstration projects, carried out in collaboration with industrial partners, and, in the case of Rio Blanco, over vigorous opposition from environmentalist and scientific advocacy groups concerned with the contamination of groundwater as well as with the health hazards associated with the continuous use of nuclear-stimulated gas. The opposition's success in the politics of precaution nevertheless came afterward when, in 1974, Colorado voters passed an amendment forbidding future nuclear explosions in the state without prior consent from the electorate.[11] With the exception of a handful of follow-up studies, Plowshare was defunded entirely after 1974.

Reflections on a Failed Technocracy

Writing in the early 1960s, the philosopher Herbert Marcuse expressed grave concern for a world in which complex moral, social, and political problems had been condensed into a narrow "technological rationality," and politics itself reduced to mere administration. Science and technology, for Marcuse, were not the engines of a universal human progress, not evidence of an increasingly perfected Enlightenment rationality, but had figured instead in the expansion of precisely the "fantastic and insane" aspects of rationality into human consciousness and public life. His was a grim vision, but hardly a baseless one at the time. Recalling the 1960 demonstration at civil defense headquarters in New York City of a "deluxe fallout-shelter" featuring "wall-to-wall carpeting ('soft'), lounge chairs, television, and Scrabble, 'designed as a combination family room during peacetime . . . and family fallout shelter should war break out,'" Marcuse warned, "if the horror of such realizations does not penetrate into consciousness, if it is readily taken for granted, it is because these achievements are . . . perfectly rational in terms of the existing order."[12]

While the space of the bomb shelter, dispersed across the American Cold War landscape, represented one solution to the anxieties of a nuclear age, the utopian vision of reshaping the earth's surface to suit human needs with nuclear explosions suggested quite another, a concerted effort on the part of the Plowshare program's "true believers", as Livermore's first director, Herbert York, reflected in a 2004 interview, to turn the bomb into something *good*.[13] The moral mission of geographical engineering was one of human control over nuclear technology as over nature itself. At the same time, of course, Plowshare was a research program developed by an enterprising weapons laboratory seeking both to expand and diversify its explosives programs and to provide a positive story of the bomb at the precise moment when fears of nuclear war, and of the increasingly tangible effects of radioactive fallout, had become palpable in calls for nuclear testing bans and limits. But were either of these technological utopias—the model family room/fallout shelter or the prospect of geographical engineering—so easily taken for granted? And how entrenched were the "existing orders," as Marcuse put it, from which the production of such spaces made sense?

This book has provided a study of the rationalities and discourses through which the enterprise of nuclear earthmoving was pursued, and an examination of how they were constituted in the collective work of scientists and engineers, technicians and technocrats, politicians, journalists, and others. But in following these actors, we have also had to examine the nature of the obstacles that they ran up against, resistance that emerged at varied social sites. Although Plowshare may have succeeded, for a time, as a utopian narrative, the ultimate failure of the geographical engineers to "change the earth's surface to suit us" was itself the result of a great deal of scientific and political work, including the production of new knowledge that could be used to call AEC and Livermore claims of safety and feasibility into question.

Following the 1966 publication of *Environment of the Cape Thompson Region, Alaska*, a pathbreaking volume in cold regions sciences based on the Project Chariot preshot environmental studies, AEC ecologist John Wolfe, still chief of the Division of Biology and Medicine's Environmental Sciences Branch, sent a copy of the 1,200-page tome to Gerald Johnson, Plowshare's first program leader who by then had become director of Navy Laboratories. Four years after the AEC had elected to hold the Chariot experiment "in abeyance," Wolfe still ruminated over the unrealized dream of geographical engineering in the Arctic. "I'm sorry no shot was ever made," he volunteered. "Such events will occur in time to come as a matter of course. This was a golden opportunity. I'm not blaming anyone, but somebody missed a boat." Johnson replied two weeks later with considerably less confidence. "I have read your Book with great interest and find it most illuminating. This perhaps will mark the only concrete result of the Plowshare Program, but I think it amply justifies it. I agree that the shot should have been fired and in 1960," Johnson added, "but . . . the water, as well as the dam, is obviously gone."[14]

By 1970 Johnson had moved to the private sector as the new manager of explosives engineering services for the subsidiary Gulf General Atomic. While the physicist still lobbied for the use of nuclear bombs to unlock fossil-fuel riches from beneath the Rocky Mountains, he had by this time come to accept the demise of the nuclear excavation program as a result of two different kinds of progress. As Johnson described it for readers of the *Bulletin of the Atomic Scientists*, "despite great accomplishments, the program to reduce radioactivity has not progressed at a rate to keep up with the rate of growth of constraints. As a consequence, nuclear excavation is probably less acceptable now than it was ten years ago."[15] The record of geographical forays and retreats traced by the nuclear earthmoving program bears out Johnson's point: Livermore indeed seemed never closer to achieving geographical engineering, not counting the great holes punched into the desert floor of the test site, than when Johnson and his colleagues endeavored to blast the experimental harbor on Ogotoruk Creek in Alaska. But what kind of progress was this? If nuclear excavation in 1970 was actually further from applied practice than it had been a decade earlier, before ten years of expensive and at times hazardous research, then perhaps General Groves's "expended knowledge" typo makes a good deal of sense. Or, as the more disturbing subtext to Johnson's point suggests, Plowshare scientists, for their part, had *always* been confident of their capabilities, and confident in the validity of extant permissible dose limits and "safety feasibility" factors, however dynamic those limits had proved to be and however unpredictable the history of nuclear testing. Livermore and AEC scientists were as a matter of practical philosophy willing to experiment with the "best available" explosives, and would likely have continued to carry out experiments and projects (or both at once) to the extent their risks could be defined, according to their own, largely self-validated standards, as acceptable. What is more, if Johnson's reflections with Wolfe about the desirability of blasting the Chariot instant harbor in 1960 are to be believed, they were willing to do so even in retrospect. Such unwillingness to question their own confidence, an unwillingness to learn, was indeed the greatest hubris of the geographical engineers.

Why did the nuclear excavation program fail to live up to its expectations? Johnson's explanation remains compelling, in all its brevity: the rate of progress in reducing radioactivity did not keep up with the "rate of growth of constraints."[16] Always, for Livermore and AEC Plowshare scientists, these constraints were seen as external to the real work of the program, that is, as outside political circumstances impinging on them and preventing the kind of progress in nuclear excavation and other programs that Plowshare scientists never seem to have lost faith in, at least while the program was active. For others, however, the increased rate in the growth of constraints was itself a matter of scientific and technological progress, forged through moments of controversy, contingency, and conflict. The story of nuclear earthmoving remains, in some ways, an anachronism, rooted in the historic spaces of a Cold War nuclear weapons laboratory. And yet the kinds of predicaments that beleaguered Plowshare—contests over whose authority or dissent counted, and where it counted, in conflicts concerning the definition of human health risks, biological costs, or technological choices and in debates over how scientific controversy and uncertainty should be understood, and at times politicized, by experts and public alike—have become characteristic problems in contemporary environmental science and politics.[17]

While Livermore and the AEC struggled to maintain their program as a "technocracy of hazards" in which, as the sociologist Ulrich Beck describes it, a monopoly on the interpretation of risk is passed "to those who caused it, of all people," this imposition of authority did not go unchallenged—in the end, Plowshare offers a window onto how such arrangements have, under certain conditions, been undermined.[18] More than a failed technology, that is, Plowshare was a failed technocracy in which Livermore and AEC scientists proved unable to *reduce* nuclear excavation to a matter of administration by technical experts. And in this sense, whatever diminishing returns the program might have produced in propaganda value, its failure was a triumph of dissent in the politics of science, technology, and environments. Opposition to Plowshare was part of the emergence of a wider environmental movement with a growing consciousness of poisons already existing in the air, water, food, and land. This movement, with its basis in the work of science and scientists, demanded the right to consider the consequences of environmental transformations before they took place, and the rights to open and free information and independent expert review as a means of assessing those consequences. Even though the triumph that I describe may have amounted, most immediately, to merely a defense of the status quo—a world *without* nuclear harbors and canals—at times it is precisely through such defenses, on the margins of possibility, that new political subjects have taken shape, mobilized around the politics of knowledge, risk, and place. Forged in contestation and controversy, new rights demands become legitimate, including demands for what I have called the rights to knowledge about the nature of ecological and health risks, and rights to information about the uncertainty of that knowledge and about the economies through which it is produced.

Marcuse's fears of technocracy were quite legitimate, of course, and have in some ways been realized. We *live* in the world of the Cold War scientific and technological project, a world of satellites and computer cartographies, electronics and the Internet, radionuclides, guided missiles and nuclear proliferation. It is a world never fully under control or predictable in which we depend on the knowledge and, at times, on the politics and governance of scientists and other experts. That our world contains few remnants of Livermore's great experiment in geographical engineering in the landscape today affirms the continuing need for a critical public science engaged with all manner of technological choices and risks, and for movements critically engaged with science and technology as well.

Epitaph

As the Atomic Energy commissioner and Nobel Prize–winning chemist Willard Libby had confidently described the capabilities of the nuclear excavation program in Geneva in 1958: "There is no doubt at all that we could make a harbor, and that we could make it at an expense which will make it very attractive in certain areas of the world."[19] By 1969, however, when the Western Australian harbor feasibility study was abruptly canceled, the certainties with which Libby spoke had been seriously called into question. Even on the coast of one of the world's more sparsely populated regions, with a seemingly friendly local booster community in place, it had been made clear that the very criteria on which Libby's essential claim "we could make a harbor" was to be evaluated, and by whom, had been transformed. If *making* a harbor meant that it was to be made safely, then who was to decide what counted as safe, and by what definition? How was this authority to be politically constituted, to what extent did it depend on scientific and technical expertise, and to what extent did it depend on open and accountable scientific and political discourse?

Livermore and the AEC's presumed authority had been increasingly undermined as Plowshare scientists attempted to take their geographical engineering techniques from the laboratory to the world outside, from one proving ground to another between 1957 and 1973. To be sure, the limits imposed on Plowshare, and the ultimate failure of the scientists to complete their ambitious experimental program, should not be celebrated to the extent that they obscure the broader, more destructive history of the American nuclear complex. And yet because the changed political conditions that Plowshare encountered were produced reciprocally with changed public knowledge and awareness of risks, it remains an important story to remember. And while the idea of engineering geographies with nuclear explosives may be dead, many different kinds of environmental transformations are presently under way, with many more possible biological, chemical, and genetic proving grounds to come in which the future remains very much at stake.

I began this book by raising questions about the failure of Plowshare's scientists and program builders to make a fundamentally geographical move, a move to take their capability of blasting big holes in the ground from one experimental space to another. It is evident after close historical study that the story of the nuclear earthmoving program is unknowable without understanding this spatiality: from the physical movement of its dust clouds to the mobility of its scientists and the mobility of their experiments, discourses, and claims to authority. In mapping the work of scientists, engineers, advisory committees, public information specialists, journalists, booster communities, state and local state institutions, and many other actors who participated in—or worked in opposition to—the making and working of Plowshare's proving grounds, I have pushed the divisions among them, and the points of controversy over which they argued, to the foreground of my analysis. It was through these controversies and oppositional relationships that Plowshare was undermined: the mobility and spatial strategies of the geographical engineers both reflected and reproduced territorialities of scientific authority that no amount of faith (among those wishing to beat their swords into plowshares, but to keep the swords too) could overcome.

What shall we write over Plowshare's grave?[20] Perhaps the final irony, for the weapons scientists who wished to experiment with and engineer geography, was a geography lesson. The very idea of geographical engineering called for the treatment of geography as an object-to-be-engineered. An Arctic creek was to be blasted into a harbor, desert mountains into a highway and railroad pass, and a tropical jungle into a canal, to mention just a few such transformations. But by focusing on the geography *of* geographical engineering, in a critical and historical sense, we have gained a quite different vision of the geography of Plowshare. Geography provides us with a dynamic spatial frame for exploring processes of boundary making, spatial and scalar relationships, and territorializations; and at the same time, it remains a place, an engineered space, and a site of social contest. The story of Plowshare, with its grandiose earthmoving schemes now confined to the cratered landscape of the Nevada Test Site and, for some of its longer-living radionuclides, to places and bodies downwind, shows precisely how science is grounded in geography in this sense, a fragmented geography from which politics and emotions could not be excluded.

Today the Sedan Crater is the leading attraction on occasional public tours of the Nevada Test Site, even though external radiation surrounding the crater remains above normal background levels. After giving a social security number to the Department of Energy's Nevada Office some weeks in advance, and, passing through military police checkpoints at the boundary of the Test Site (and, later, at a checkpoint inside the grounds), American citizens may be invited to a viewing platform at the edge of the crater to witness the most striking evidence of the scale of power that the Plowshare scientists hoped to put to work in their earthmoving projects. But far from the beginning of a massive program of atomic age geographical engineering for which the Crater had been intended, nuclear excavation under Plowshare actually peaked with Sedan in 1962; in fact, more than forty years later, it remains not only the largest Plowshare crater, but the most massive of all the nearly one hundred surface and near-surface explosions set off at the Nevada Test Site between 1951 and 1992. The crater itself now stands as a fitting epitaph to Plowshare, to beliefs in unlimited control over nature and technology, to nuclear testing practices guided by haste and arrogance, and, not least, to the limits imposed on these practices. Here lies Plowshare, because it lies nowhere else.

Notes

Abbreviations

AEC	United States Atomic Energy Commission
DOE	United States Department of Energy
JCAE	Joint Committee on Atomic Energy, U.S. Congress
LLNL	University of California Lawrence Livermore National Laboratory
LTBT	Treaty Banning Nuclear Weapons Tests in the Atmosphere, in Outer Space, and Underwater (Limited Test Ban Treaty)
NTA	U.S. Department of Energy Nuclear Test Archive, Las Vegas, Nevada

Introduction: Geographical Engineering

1. Public Law 88–609, 88th Congress, S. 2701, September 22, 1964.
2. E. Graves, U.S. Army Corps of Engineers Nuclear Cratering Group, "Nuclear Excavation of a Sea-Level, Isthmian Canal," *Proceedings of the Third Plowshare Symposium*, April 1964, Plowshare Box P-26, Office of History and Historical Records, LLNL; L. J. Vortman, "Construction of a Sea-Level, Trans-Isthmian Canal Using Nuclear Explosives," SC-4929(RR), February 1964, Plowshare Box P-24, LLNL; L. J. Vortman to distribution, "Transcontinental Canal," August 26, 1964, Plowshare Box P-24, LLNL.
3. Graves, "Nuclear Excavation."
4. Edward Teller (with Allen Brown), *The Legacy of Hiroshima* (Garden City, N.Y.: Doubleday and Company, 1962), 84.
5. Edward Teller, Wilson K. Talley, Gary Higgins, and Gerald Johnson, *The Constructive Uses of Nuclear Explosives* (New York: McGraw-Hill Book Company, 1968), vi.
6. This book builds on a number of important but partial studies of Plowshare's nuclear excavation program, including Trevor Findlay, *Nuclear Dynamite: The Peaceful Nuclear Explosions Fiasco* (Sydney: Brassey's Australia, 1990), an international survey of peaceful nuclear explosives programs and their implications for nuclear testing regimes; Dan O'Neill, *The Firecracker Boys* (New York: St. Martin's Press, 1994), a history of the Project Chariot controversy in Alaska; and Scott Kirsch, ed., "Nuclear Engineering and Geography" theme issue of *Ecumene* 5 (1998): 263–322.
7. Bruno Latour, *Science in Action: How to Follow Scientists and Engineers through Society* (Cambridge, Mass.: Harvard University Press, 1987).
8. Kevin O'Neill, "Building the Bomb," in *Atomic Audit: The Costs and Consequences of U.S. Nuclear Weapons since 1940*, ed. S. Schwartz (Washington, D.C.: Brookings Institution Press, 1998), 33–104.

9. Milo Nordyke, "Discussion of the Need for a Harbor Demonstration," UOPKC 68–106, November 18, 1968, Plowshare Box P-22, LLNL.
10. Several underground Plowshare tests in natural gas stimulation were carried out (in New Mexico in 1967 and Colorado in 1969 and 1973). The events of these commercial endeavors are beyond the scope of this book, but will be discussed in subsequent chapters to the extent that they affected the common grounds of the Plowshare program. On the Colorado underground tests, see F. Kreith and C. Wrenn, *The Nuclear Impact* (Boulder, Colo.: Westview Press, 1976). See also Findlay, *Nuclear Dynamite*, 181–203.
11. Steven Shapin, "The House of Experiment in Seventeenth-Century England," *Isis* 79 (1988): 373–404; Shapin and Simon Schaffer, *Leviathan and the Air Pump: Hobbes, Boyle, and the Experimental Life* (Princeton: Princeton, N.J.: University Press, 1985). See also Thomas Gieryn, *Cultural Boundaries of Science: Credibility on the Line* (Chicago: University of Chicago Press, 1999).
12. Scott Kirsch, "Peaceful Nuclear Explosions and the Geography of Scientific Authority," *Professional Geographer* 52 (2000): 179–192. For a related point, see John Cloud, "Imaging the World in a Barrel: CORONA and the Clandestine Convergence of the Earth Sciences," *Social Studies of Science* 31 (2001): 231–251.
13. Michael Heffernan, "Bringing the Desert to Bloom: French Ambitions in the Sahara Desert during the Late Nineteenth Century—the Strange Case of '*la mer intérieure*,'" in *Water, Engineering, and Landscape*, ed. D. Cosgrove and G. Petts (London: Bellhaven Press, 1990), 108.
14. Edward Teller, "University Role in Nuclear Explosives Engineering Research," in *Education for Peaceful Uses of Nuclear Explosives*, ed. Lynn E. Weaver (Tucson: University of Arizona Press, 1970), 294.

1. *Origins of a Cold War Experimental Program*

1. Edward Teller, Wilson K. Talley, Gary Higgins, and Gerald Johnson, *The Constructive Uses of Nuclear Explosives* (New York: McGraw-Hill Book Company, 1968), vi.
2. Trevor Findlay, *Nuclear Dynamite: The Peaceful Nuclear Explosions Fiasco* (Sydney: Brassey's Australia, 1990), 5. Johnson is quoted from a January 17, 1987, letter to Findlay.
3. At least one document suggests practical interest in this project at the laboratory, without reference to nuclear explosives. Eng. N. Pesakhowich, Tel Aviv, Israel, "Brief Outline of the Israel Navigation Canal Project," July 7, 1957, Plowshare Box P-23, LLNL. Edward Teller also describes a later visit to Israel in Edward Teller and Judith Shoolery, *Memoirs: A Twentieth-Century Journey in Science and Politics* (New York: Perseus Publishing, 2002), 498–499.
4. Hugh Gusterson, *Nuclear Rites: A Weapons Laboratory at the End of the Cold War* (Berkeley: University of California Press, 1996); Peter J. Westwick, *The National Labs: Science in an American System, 1947–1974* (Cambridge, Mass.: Harvard University Press, 2003).
5. In fact, Frederick Reines of the Los Alamos laboratory had explored the idea several years earlier in the technical context of an earlier generation of atomic weaponry and had essentially dismissed the notion because of the fallout problem. Reines, "Are There Peaceful Engineering Uses of Atomic Explosives?" *Bulletin of the Atomic Scientists*, June 1950, 171–172. Reines's paper was written in response to the statement by Soviet foreign minister Andrei Vishinsky in 1949, following the first Soviet atomic test, which asserted that, in the Soviet program, "We are razing mountains, we are irrigating deserts, we are cutting through the jungle, we are spreading life, happiness, prosperity and welfare in places where the human footstep has not been seen for thousands of years." Quoted in Findlay, *Nuclear Dynamite*, 2.

6. Brown to York, February 26, 1957, "Memorandum: Non-Military Uses Symposium," DOE doc. NV0134664, NTA. Early program goals are also worked out in York to Starbird, September 27, 1957, DOE doc. NV0078541, NTA; Brown and Johnson to Starbird, November 27, 1957, DOE doc. NV0134628, NTA.
7. Brown to York, February 26, 1957. Emphasis in original.
8. Ibid.
9. Ibid.
10. As late as April 1957, Atomic Energy commissioner Willard Libby complained to the AEC general manager that neither Los Alamos nor Livermore was doing substantial work in the area of nonmilitary uses. Libby was presumably aware of the February symposium, so his remarks may have been intended to expedite the AEC's institutionalization of the program. W. F. Libby, "Memorandum to the General Manager," April 2, 1957, DOE doc. NV0073525, NTA. See also York to Starbird, September 27, 1957; Brown and Johnson to Starbird, November 27, 1957. According to Teller, the name "Project Plowshare" originated when Harold Brown mentioned the nonmilitary program to physicist I. I. Rabi, who responded, "So you want to beat your old atomic bombs into plowshares," updating the well-known words from Isaiah 2: 4. Edward Teller (with Allen Brown), *The Legacy of Hiroshima* (Garden City, N.Y.: Doubleday and Company, 1962), 82.
11. Interview with Herbert York, August 4, 2004.
12. These questions are adapted from David Livingstone, "The Spaces of Knowledge: Contributions towards a Historical Geography of Science," *Society and Space* 12 (1995): 5–35.
13. Gerald D. Nash, *The Federal Landscape: An Economic History of the Twentieth Century West* (Tucson: University of Arizona Press, 1999). For additional regional historical perspectives, see B. Hevly and J. M. Findlay, eds., *The Atomic West* (Seattle: University of Washington Press, 1998); Ann Markusen, Peter Hall, Scott Campbell, and Sabina Dietrick, *The Rise of the Gunbelt: The Military Remapping of Industrial America* (New York: Oxford University Press, 1991).
14. My account here draws especially from Stanley Goldberg, "General Groves and the Atomic West: The Making and Meaning of Hanford," in *The Atomic West*, ed. Hevly and Findlay, 39–89; Peter Bacon Hales, *Atomic Spaces: Living on the Manhattan Project* (Champaign: University of Illinois Press, 1997); Richard Rhodes, *The Making of the Atomic Bomb* (New York: Simon and Schuster, 1986); Alice Kimball Smith, "Manhattan Project: The Atomic Bomb," in MIT Faculty Coalition for Disarmament, *The Nuclear Almanac: Confronting the Atom in War and Peace*, ed. J. Dennis (Reading, Mass.: Addison-Wesley Publishing Company, 1984), 21–42.
15. Many physicists had already imposed some form of censorship on themselves by this time so as not to inadvertently help a German atomic bomb program.
16. Hales, *Atomic Spaces*. On Hales's treatment of Manhattan Project science and scientists, see my review in the *Annals of the Association of American Geographers* 89 (1999): 166–168.
17. Hales, *Atomic Spaces*, 42. The primary settings for scientific research before the creation of Los Alamos included the University of Chicago, the University of California, Berkeley, and Columbia University, all of which remained the settings for Manhattan Project work after the creation of Los Alamos as well.
18. Jessica Wang, *American Science in an Age of Anxiety: Scientists, Anti-Communism, and the Cold War* (Chapel Hill: University of North Carolina Press, 1999). On the atomic scientists' movement, and its role in public debate over the May-Johnson bill and what would become the Atomic Energy Act of 1946, see also Brian Balogh, *Chain Reaction: Expert Debate and Public Participation in American Commercial Nuclear Power, 1945–1975* (Cambridge: Cambridge University Press, 1991); Paul Boyer,

By the Bomb's Early Light: American Thought and Culture at the Dawn of the Atomic Age, 2nd ed. (Chapel Hill: University of North Carolina Press, 1994); Alice Kimball Smith, *A Peril and a Hope: The Scientists' Movement in America, 1945–1947* (Chicago: University of Chicago Press, 1965).

19. For an institutional history of the GAC, see Richard Sylves, *The Nuclear Oracles: A Political History of the General Advisory Committee of the Atomic Energy Commission* (Ames: Iowa State University Press, 1987).
20. Dan O'Neill, *The Firecracker Boys* (New York: St. Martin's Press, 1994).
21. Wang, *American Science*, 7. From 1946 to 1953 the AEC investigated half a million people (including, but not limited to, scientists) for security clearances, an enterprise costing roughly ten million dollars per year. Peter J. Westwick, "In the Beginning: The Origin of Nuclear Secrecy," *Bulletin of the Atomic Scientists*, November/December 2000, 43–49. On the military redirection of research priorities in the physical sciences during the early Cold War, see Paul Forman, "Behind Quantum Electronics: National Security as Basis for Physical Research in the United States, 1940–1960," *Historical Studies in the Physical and Biological Sciences* 18 (1987): 149–229; see also Daniel Kevles, "Cold War and Hot Physics: Science, Security, and the American State, 1945–56," *Historical Studies in the Physical and Biological Sciences* 20, (pt. 2 1990): 239–264.
22. Westwick, *National Labs*.
23. This move was part of a more widespread militarization of space during World War II and the Cold War. Between 1937 and 1957, military lands in the continental United States and Alaska were enlarged tenfold, from three million to thirty million acres. Not until the passage of the Engle Act of 1958 were defense agencies required to gain congressional approval for land withdrawals greater than five thousand acres. Rather, land could be withdrawn from the public domain simply by specifying for the Department of Interior the requested area and indicating that it was vital to military operations. Myriam Weisang Misrach "Part I: The Story," in Richard Misrach with Myriam Weisang Misrach, *Bravo 20: The Bombing of the American West* (Baltimore: Johns Hopkins University Press, 1990), 1–49. See also Valerie L. Kuletz, *The Tainted Desert: Environmental and Social Ruin in the American West* (London: Routledge, 1998); David Loomis, *Combat Zoning: Military Land-Use Planning in Nevada* (Reno: University of Nevada Press, 1993); Nash, *Federal Landscape*.
24. Anamosa to McCool, "Mr. Strauss Memorandum Regarding Presidential Directive to Hold No Further Tests," September 3, 1954, Office Memorandum, United States Government, DOE doc. NV0031078, NTA; McCormack to Parsons, "Continental Proving Ground," August 25, 1948, DOE doc. NV0028604, NTA; Snapp to McCormack, "Commission Action on AEC 141—Location of Proving Ground for Atomic Weapons," September 22, 1948, DOE doc. NV0028603, NTA; Tyler to McCormack, "Selection of a Continental Proving Ground," December 1, 1950, DOE doc. NV0028595, NTA; U.S. AEC, "Atomic Energy Commission: Location of Proving Ground for Atomic Weapons—Memo from National Security Council; note by the acting secretary," November 16, 1950, AEC 141/6, NTA; U.S. AEC, "Selection of a Continental Atomic Test Site; report by the Director of Military Application," as attached to "Atomic Energy Commission: Location of Proving Ground for Atomic Weapons," December 13, 1950, DOE doc. 30419. See also Barton Hacker, *Elements of Controversy: The Atomic Energy Commission and Radiation Safety in Nuclear Weapons Testing, 1947–1974* (Berkeley: University of California Press, 1994). Parts of this account of the origins and early testing activities of the Nevada Test Site appeared, in an earlier form, in Scott Kirsch, "Watching the Bombs Go Off: Photography, Nuclear Landscapes, and Spectator Democracy," *Antipode* 29 (1997): 227–255.
25. U.S. AEC, "Location of Proving Ground for Atomic Weapons."

26. Anamosa to McCool, "Mr. Strauss Memorandum."
27. U.S. AEC, "Selection of a Continental Atomic Test Site."
28. Ibid.
29. Ibid.; Hacker, *Elements of Controversy*, 40–44. An assessment of safety in Nevada was also made by Dr. Charles Dunham of the AEC's Division of Biology and Medicine, who insisted that no serious problems should be expected, save perhaps for the town of Alamo, located fifty miles downwind from the site, which he suggested could be easily evacuated if necessary. Kevin O'Neill, "Building the Bomb" in *Atomic Audit: The Costs and Consequences of U.S. Nuclear Weapons since 1940*, ed. S. Schwartz (Washington, D.C.: Brookings Institution Press, 1998), 33–104, 51.
30. U.S. AEC, press release, January 11, 1951. DOE doc. NV0033120, NTA.
31. C. P. Trussell, "Atom Bomb Testing Ground Will Be Created in Nevada," *New York Times*, January 12, 1952, 1, 12.
32. Teller's thesis dealt with calculating energy levels of hydrogen molecules in an excited state. After moving to the United States, his research focus shifted from the molecular to the atomic level, and soon to the release of energy from nuclear and thermonuclear explosions. Teller's more or less authorized biographies include Stanley A. Blumberg and Gwinn Owens, *Energy and Conflict: The Life and Times of Edward Teller* (New York: G. P. Putnam's Sons, 1976), and Stanley A. Blumberg and Louis G. Panos, *Edward Teller: Giant of the Golden Age of Physics* (New York: Charles Scribner's Sons, 1990). See also William J. Broad, *Teller's War* (New York: Simon and Schuster, 1992), and Herbert York, *The Advisors: Oppenheimer, Teller, and the Superbomb* (San Francisco: W. H. Freeman and Company, 1976). Perhaps the most compelling source on Teller so far, though, is Teller, *Legacy of Hiroshima*; see also Teller and Shoolery, *Memoirs*.
33. Blumberg and Owens, *Energy and Conflict*, 119.
34. Cf. Teller, *Legacy of Hiroshima*; Edward Teller, *The Reluctant Revolutionary* (Columbia: University of Missouri Press, 1964).
35. The Teller-Ulam H-bomb was a design by which "a relatively small fission explosion can induce an arbitrarily large thermonuclear explosion." Herbert York, "The Origins of the Lawrence Livermore Laboratory," *Bulletin of the Atomic Scientists*, September 1975, 8–14, 11.
36. York, "Origins"; Teller, *Legacy of Hiroshima*.
37. York, "Origins," 12.
38. Livermore has undergone periodic name changes. Following Lawrence's death in 1958, the laboratory was renamed the Lawrence Radiation Laboratory at Livermore, and in 1971 it became the Lawrence Livermore National Laboratory, still managed by the University of California. I refer to then-current laboratory names, or the shorthand "Livermore," throughout my account.
39. York, "Origins," 13.
40. Westwick, *National Labs*.
41. All figures in current dollars. Westwick, *National Labs*; Gusterson, *Nuclear Rites* (employment figures).
42. Stuart W. Leslie, *The Cold War and American Science: The Military-Industrial-Academic Complex at MIT and Stanford* (New York: Columbia University Press, 1993); see also Noam Chomsky et al., *The Cold War and the University: Toward an Intellectual History of the Postwar Years* (New York: New Press, 1997); Peter Galison and Bruce Hevly, eds., *Big Science: The Growth of Large-Scale Research* (Stanford, Calif.: Stanford University Press, 1992).
43. Sherman quoted in U.S. AEC, "Admission of Limited, Selected Group of Uncleared Observers to One Shot Spring 1952 Test Series in Nevada," March 11, 1952, AEC 505/15. DOE doc. NV0072644, NTA.

44. "AEC v. the Reporters," *Time*, November 12, 1951, 42.
45. U.S. AEC, "Tumbler-Snapper Information Plan," U.S. AEC Santa Fe Operations Office, DOE doc. NV0026272, NTA.
46. U.S. AEC, "Admission of Limited, Selected Group of Uncleared Observers."
47. The Atomic Energy Act of 1954 was created partly to smooth the way for private companies to get involved in the nuclear power industry (nuclear energy, and not peaceful nuclear explosions, is what Eisenhower meant by "Atoms for Peace"), and also to fill some of the legal gaps left from the Atomic Energy Act of 1946. The Atoms for Peace program also promoted the distribution of radioisotopes for use in scientific research.
48. For more detailed studies of fallout controversies during this period, see Howard Ball, *Justice Downwind: America's Atomic Testing Program in the 1950s* (New York: Oxford University Press, 1986); Phillip Fradkin, *Fallout: An American Nuclear Tragedy* (Tucson: University of Arizona Press, 1989); J. Fuller, *The Day We Bombed Utah: America's Most Lethal Secret* (New York: New American Library, 1984); Peter Goin, *Nuclear Landscapes* (Baltimore: Johns Hopkins University Press, 1991); Hacker, *Elements of Controversy*; S. Hilgartner, R. Bell, and R. O'Connor, *Nukespeak: Nuclear Language, Visions, and Mindset* (San Francisco: Sierra Club Books, 1982); A. Constandina Titus, *Bombs in the Backyard: Atomic Testing and American Politics*, 2nd ed. (Reno: University of Nevada Press, 2001).
49. See U.S. AEC, "Some Thoughts on Radiation Protection Criteria at the Nevada Test Site," n.d., DOE doc. NV0068015, NTA.
50. Bradbury to Tyler, January 5, 1953, included in U.S. AEC, "Report on the Operational Future of the Nevada Proving Grounds," U.S. AEC, Santa Fe Operations Office, 1953, DOE doc. NV0128723, NTA.
51. Lewis Strauss, "Statement by Lewis Strauss, Chairman, Atomic Energy Commission, March 31, 1954," *Bulletin of the Atomic Scientists* 10 (1954): 163–164.
52. See Barton Hacker, "'Hotter than a $2 Pistol': Fallout, Sheep, and the Atomic Energy Commission, 1953–1986," in *The Atomic West*, ed. Hevly and Findlay; Hilgartner, Bell, and O'Connor, *Nukespeak*.
53. Strauss, "Statement."
54. For a useful summary of the debate, see Carolyn Kopp, "The Origins of the American Scientific Debate over Fallout Hazards," *Social Studies of Science* 9 (1979): 403–422. The Division of Biology and Medicine was established under the AEC in 1948 to support research in the life sciences and biomedicine.
55. As Sturtevant argued—partly in response to Strauss's statement—in his 1954 presidential address to the Pacific Division of the American Association for the Advancement of Science, these observations were "so widely confirmed that we may confidently assert that they apply to all higher organisms including man." Quoted in Kopp, "Origins," 405–406.
56. "Nuclear Weapons Tests" *Science* 124 (November 9, 1956): 925–926.
57. Ibid. See also Elof Axel Carlson, *Genes, Radiation, and Society: The Life and Work of H. J. Muller* (Ithaca, N.Y.: Cornell University Press, 1981), 368–413.
58. E. B. Lewis, "Leukemia and Ionizing Radiation," *Science* 125 (May 17, 1957): 965–972.
59. U.S. AEC, *Atomic Tests in Nevada* (Washington, D.C.: U.S. Government Printing Office, 1957), 15.
60. Interview with Barry Commoner, July 29, 2004. CNI, to give one striking example, even enlisted the public as *data collectors* in a scientific project, a campaign to collect 50,000 baby teeth to investigate how children's bodies, in particular, were susceptible to bone-seeking radionuclides like Sr-90. W. Wyant, "50,000 baby teeth," *The Nation*, June 13, 1959. See also Barry Commoner, *Science and Survival* (New York:

Viking Press, 1963); M. Katz, *A History of SANE, the Committee for a Sane Nuclear Policy, 1957–1985* (New York: Greenwood Press, 1986); F. McCrea and G. Markle, *Minutes to Midnight: Nuclear Weapons Protest in America* (Newbury Park, Calif.: Sage Publications, 1989).

61. Katherine Magraw, "Teller and the 'Clean Bomb' Episode," *Bulletin of the Atomic Scientists*, May 1988, 32–37. Magraw makes an interesting contrast between the "clean bomb" episode and the "Star Wars" episode of 1987.
62. Ibid., 32. Research on the "clean bomb"—known later as the neutron bomb—sought to minimize the fission-to-fusion ratio. Its potential value, of course, was seen not only in peaceful explosions but in creating new opportunities for so-called limited nuclear wars. On the advice of Teller, Lawrence, Mills, and Strauss, Eisenhower latched onto clean-bomb research as a humane alternative, especially in the event of nuclear war, that the country could not turn its back on in the research stage.
63. Between May 27, 1957, and October 30, 1958, as negotiations for a moratorium or other possible treaties were carried on, the AEC tested 108 atomic explosives (including 70 in Nevada, 35 in the Marshall Islands and other Pacific proving grounds, and 3 missile tests in the South Atlantic). U.S. Department of Energy, *United States Nuclear Tests: July 1945 through September 1992*, DOE/NV-209 (Rev. 14), December 1994.
64. Quoted in Linus Pauling, letter to the editor, *Life*, March 17, 1958, 21.
65. Latter was a theoretical physicist who had also worked with Teller on the air force's Scientific Advisory Board. Edward Teller and Albert Latter, *Our Nuclear Future: . . . Facts, Dangers, and Opportunities* (New York: Criterion Books, 1958); Edward Teller and Albert Latter, "The Compelling Need for Nuclear Tests," *Life*, 19 February 1958, 64–72.
66. Teller and Latter, "Compelling Need for Nuclear Tests," 65.
67. A roentgen is the international unit of measurement for X-radiation or gamma radiation, signifying one electrostatic unit of ionizing charge.
68. Teller and Latter, "Compelling Need for Nuclear Tests."
69. Pauling, letter to the editor; Edward Teller and Albert Latter, "Doctors Teller and Latter Reply," *Life*, March 17, 1958, 21.
70. Linus Pauling, "Fact and Fable of Fallout," *Nation*, July 14, 1958, 537–542. See also Linus Pauling, *No More War!* (New York: Dodd, Mead, 1958).
71. Pauling, letter to the editor.
72. Teller and Latter, "Doctors Teller and Latter Reply."
73. Edward Teller with Allen Brown, "The Real Dangers of Radiation," *This Week Magazine*, October 9, 1960, 7–11.
74. "Nuclear weapons tests."
75. Thomas Gieryn, *Cultural Boundaries of Science: Credibility on the Line* (Chicago: University of Chicago Press, 1999).
76. Teller, *Legacy*, 84.
77. Jim Frank, Jim Ruff, and Harlan Zodtner to Distribution, "Memorandum: Glossary of Terms for Plowshare Program," September 18, 1958, Plowshare Box P–29, LLNL.
78. W. F. Libby, "Remarks Prepared by Dr. Willard F. Libby, Commissioner, United States Atomic Energy Commission, for Delivery at Amherst College, Amherst Massachusetts, April 30, 1958," U.S. AEC press release, April 30, 1958, DOE doc. NV0005046, NTA.
79. "Peaceful Atomic Blasting," *Time*, March 24, 1958, 64.
80. Neal Stanford, "Dr. Libby's Earth-Shaking Story," *Foreign Policy Bulletin*, April 15, 1958, 115.
81. Libby, "Remarks."
82. Alden P. Armagnac, "Atomic Blasting for Peacetime Feats," *Popular Science*, September 1958, 102–106.

83. Gerald W. Johnson and Harold Brown, "Non-Military Uses of Nuclear Explosives" *Scientific American* 199 (December 1958): 29–36, 33. See also H. Brown and G. W. Johnson, "Non-Military Uses of Nuclear Explosions," UCRL 5026, University of California Radiation Laboratory, June 12, 1958.
84. Teller, *Legacy of Hiroshima*, 92, 79, 90; Edward Teller, "We're Going to Work Miracles," *Popular Mechanics* 113 (March 1960): 97–101, 278–282.
85. O'Neill, *Firecracker Boys*.
86. John W. Finney, "U.S. Atom Experts Wary over Soviet," *New York Times*, September, 12, 1958, 7.
87. Ibid.

2. Toward an "Early and Obvious Demonstration"

1. York to Starbird, September 27, 1957, DOE doc. NV0078541, NTA.
2. Brown to York, February 26, 1957, "Memorandum: Non-Military Uses Symposium," DOE doc. NV0134664, NTA.
3. Brown and Johnson to Starbird, November 27, 1957, DOE doc. NV0134628, NTA. To propose the experiment, it was also to propose a substantial increase in funding for the Plowshare program, since actual nuclear tests and their related experiments subsumed the lion's share of costs for laboratory programs. Indeed, Plowshare funding was increased from a nominal $100,000 in its first fiscal year to $3,000,000 for the FY 1959 program.
4. The language of demonstration has a long history of its own in the experimental sciences in connection with movements across public and private boundaries. See Steven Shapin, "The House of Experiment in Seventeenth-Century England," *Isis* 79 (1988): 373–404.
5. Brown and Johnson to Starbird, November 27, 1957.
6. Ibid.; Teller to Starbird, August 15, 1958, Plowshare Box P-49, LLNL.
7. H. H. Zodtner, "An Investigation of the feasibility of building a harbor on the west coast of South America, using explosive power of Nuclear Weapons: a Preliminary Report," January 17, 1958, Plowshare Box P-25, LLNL.
8. Teller to Fidler, October 7, 1958, DOE doc. NVO125233, NTA.
9. E. J. Longyear Company, "Report to University of California Radiation Laboratory on the Mineral Potential and Proposed Harbor Locations in Northwestern Alaska," April 18, 1958, Gerald Johnson Papers, MSS. 206, Box 9, Folder 13, Mandeville Special Collections Library, University of California, San Diego; University of California Radiation Laboratory (Livermore) purchase order 1934807, quoted in E. J. Longyear Company, "Report."
10. Atomic Energy Commission meeting 1369, May 6, 1958, DOE doc. NV0001171, NTA; Bacigalupi to Kilgore, May 20, 1958, Plowshare Box P-49, LLNL. The self-identified general term for Alaskan "Eskimos" today is Inupiat or Inupiaq. The Inupiat are further subdivided by village or community, for example, the Tigaraqmiut at Point Hope, the Naupaktomiut and Noatagmiut at Noatak, and the Kivalingmiut at Kivalina. I will generally use the term Inupiat, but for narrative consistency, will also at times use the term Eskimo, which is what the Inupiat were generally called by the primary informants of this analysis, the Chariot environmental scientists, and also commonly by the people of Point Hope themselves during this period.
11. U.S. AEC press release, June 9, 1958, San Francisco Operations Office, Don C. Foote Collection, Box 11, Folder 18, Archives, University of Alaska Fairbanks. See also *New York Times*, "U.S. May Use A-Bombs to Make Alaska Harbor," June 10, 1958; AEC 811/37, March 3, 1959, reference to meeting 1381, June 12, 1958. DOE

doc. NV0070391, NTA. See also Atomic Energy Commission meeting 1369, May 6, 1958, DOE doc. NV0001171, NTA.

12. For a more detailed account of this and subsequent visits of Livermore scientists to Alaska in connection with Project Chariot, see Dan O'Neill, *The Firecracker Boys* (New York: St. Martin's Press, 1994). See also Scott Kirsch and Don Mitchell, "Earth-moving as the 'Measure of Man': Edward Teller, Geographical Engineering, and the Matter of Progress," *Social Text* 54 (1998): 100–134.
13. "Dr. Teller and Team to Study Use of Nuclear Power Here," *Juneau Daily Empire*, July 15, 1958.
14. E. J. Longyear Company, "Report." This railroad estimate exceeds AEC figures, which purported that 50–100 miles of railroad would be enough. O'Neill, *Firecracker Boys*.
15. "Alaska Can Have Massive Nuclear Engineering Job," *Fairbanks Daily News-Miner*, July 17, 1958, cited in O'Neill, *Firecracker Boys*, 37.
16. The field conference at Cape Thompson took place June 17–19, 1958. Teller to Shute, January 26, 1959. Included in AEC 811/37, Atomic Energy Commission: Proposed Revisions to the Program for Peaceful Uses of Nuclear Explosives, March 3, 1959, DOE doc. NV0070391, NTA. Holmes and Narver was a Livermore contracting firm involved in engineering, construction, and support operations.
17. One might plausibly argue, given laboratory director Teller's stated positions, that the choice of the Alaskan location near Siberia was actually intended to disrupt the test moratorium and future treaty negotiations, although there is no evidence to support this view. More likely, I would argue, the optimism among Livermore scientists that fallout could be controlled rendered the proximity of the Soviet Union, in their view, an insignificant problem as long as fallout and debris were sent into weather blowing eastward over Alaska.
18. Minutes, Fifty-ninth Meeting of the General Advisory Committee to the United States Atomic Energy Commission, July 21, 22, and 23, 1958, Berkeley and Livermore, Calif., October 22, 1958, DOE doc. NV0073720, NTA.
19. Ibid. This proposed yield was soon revised downward to a total of 2.4 megatons, still 160 times the power of Hiroshima.
20. Teller to Starbird, August 15, 1958.
21. Teller to Fiddler, October 7, 1958. Although there was, as O'Neill indicates in *The Firecracker Boys*, considerable confusion among AEC commissioners in early 1959 concerning whether Project Chariot was still viable, there seem never to have been any doubts at Livermore, where the idea of revising the experiment downward in scale had taken root by August 1958.
22. Ibid.
23. Teller to Shute, January 26, 1959. Studies exploring the possibility of a nuclear-excavated, sea-level Central American canal were under way by this time and were pursued actively throughout the year. Report UCRL-5253, "Industrial Uses of Nuclear Explosives, Plowshare Series Report #1," Plowshare Box P-34, LLNL; L. J. Vortman, "Excavation of a Sea-Level Ship Canal," in *Proceedings of the Second Plowshare Symposium, May 13–15, 1959*, San Francisco, California, UCRL-5676, DOE doc. NV0015962, NTA. Meetings with the Army Corps of Engineers and other affiliated agencies were initiated in 1959: cf. "Attendees Panama Meetings" and "Outline," September 24 and 25, 1959, Plowshare Box P-24, LLNL; "Trans-Isthmian Canal," October 30, 1959, Plowshare Box P-24, LLNL; Teller to Potter telex, December 14, 1959, Plowshare Box P-24, LLNL.
24. Teller to Shute, January 26, 1959. Estimated cost for the 460-kiloton (kt) experiment was scaled back to $2,773,000, which did not include any actual harbor facilities. Soon the explosive yield for Chariot would be revised downward again

to 280 kt, with a single 200-kt turning basin and a "row-charge" of four 20-kt explosions.

25. AEC 811/37, Atomic Energy Commission: Proposed Revisions to the Program for Peaceful Uses of Nuclear Explosives, March 3, 1959, DOE doc. NV0070391, NTA. Emphasis added. Final authority for any nuclear test (or test series) rested with the president.
26. "Atom Scientists Declare Harbor Can Safely Be Blasted at Cape Thompson," *Anchorage Daily News*, January 7, 1959; "Experts Seek Alaskan Ideas on Blasting Atomic Harbor," *Fairbanks Daily News-Miner*, January 9, 1959.
27. "Atomic Harbor OK Vital," *Fairbanks Daily News-Miner*, January 10, 1959.
28. *Fairbanks Daily News-Miner*, February 9, 1959.
29. Statement read by English to Keller and Shelton, January 9, 1959, Project Chariot Collection, Box 3, Folder 20, Archives, University of Alaska Fairbanks.
30. English to Johnson, January 21, 1959, Project Chariot Collection, Box 3, Folder 20, Archives, University of Alaska Fairbanks.
31. Stephen Bocking, *Ecologists and Environmental Politics: A History of Contemporary Ecology* (New Haven: Yale University Press, 1997); Joel B. Hagen, *An Entangled Bank: The Origins of Ecosystem Ecology* (New Brunswick, N.J.: Rutgers University Press, 1992).
32. Hagen, *Entangled Bank*, 100; Eugene Odum, *Fundamentals of Ecology*, 3rd ed. (Philadelphia: Saunders, 1971), 6, 467.
33. Seventieth Meeting of the Advisory Committee for Biology and Medicine, U.S. Atomic Energy Commission, October 17–18, 1958, Germantown, Maryland, DOE doc. NV0000898, NTA.
34. Ibid. Emphasis added.
35. Johnson to English, January 26, 1959, Project Chariot Collection, Box 3, Folder 20, Archives, University of Alaska Fairbanks.
36. Patty to Wiegman and others, February 18, 1959, Project Chariot Collection, Box 3, Folder 20, Archives, University of Alaska Fairbanks.
37. O'Neill, *Firecracker Boys*.
38. Brina Kessel, Albert Johnson, Gerald Swartz, and William Pruitt, "An Ecological Study of the Flora and Fauna of the Cape Thompson-Ogotoruk Creek Region, Alaska," proposal submitted to the Atomic Energy Commission, Division of Biology and Medicine, March, 1959, Project Chariot Collection, Box 9, Folder 74, Archives, University of Alaska Fairbanks.
39. Susan Millar and Don Mitchell, "Spectacular Failure, Contested Success: the Project Chariot Bioenvironmental Programme," *Ecumene* 5 (1998): 287–302.
40. For example, see U.S. AEC press releases, March 6 and June 24, 1959, Don C. Foote Collection, Box 11, Folder 18, Archives, University of Alaska Fairbanks.
41. Quoted in O'Neill, *Firecracker Boys*, 82.
42. U.S. AEC, San Francisco Operations Office, "Chariot Environmental Program," June 20, 1959, DOE doc. NV0118055, NTA.
43. Ibid. This work, as carried out by the geographer Don Foote, will be discussed in detail in the next chapter.
44. "Charter: Committee on Environmental Studies for Project Chariot, Plowshare Program," Appendix A in "Chariot Environmental Program," U.S. AEC, San Francisco Operations Office, July 1959. SAN refers to the San Francisco Operations Office. Announced to Wolfe's committee in the June 24, 1959, press release were: Dr. Allyn Seymour, Laboratory of Radiation Biology, University of Washington; Dr. Norman Wilimovsky, Ichthyology, Bureau of Commercial Fisheries, U.S. Fish and Wildlife Service, Juneau; Dr. Robert Rausch, Arctic Health Research Center, U.S. Public Health Service, Anchorage; Dr. Kermit H. Larson, chief, En-

vironmental Radiation Division, Laboratories of Nuclear Medicine and Radiation Biology, University of California at Los Angeles; Dr. Maxwell Britton, Geography Branch (Arctic Research), Office of Naval Research, Washington, D.C.; and Ernest Campbell, AEC San Francisco Operations Office. Campbell was secretary for the committee. Larson and Seymour both relied on AEC funding for their research prior to Chariot.

45. Ibid.
46. U.S. AEC press release, June 24, 1959.
47. Ibid. Unlike the press release announcing Project Chariot a year earlier (see note 11 above), the initial press releases announcing the Chariot environmental program attracted little attention from the national media.
48. U.S. AEC press releases, March 6 and June 24, 1959; Project Manager's Operations Order 1–59, Project Chariot—Phase II, 1959, Plowshare Program, U.S. Atomic Energy Commission, Cape Thompson Area, Alaska, June 1, 1959, DOE doc. NV0018538, NTA; K. Larson, "Proposed Program, Phase I, Operation Chariot: Radiological and Biological Availability Assessment of Fallout, July 1, 1959–September 30, 1959," June 9, 1959, DOE doc. NV0026499, NTA; "Chariot Environmental Program," U.S. AEC, San Francisco Operations Office, July 1959.
49. Ibid. Other contract scientists in the environmental program included researchers in radiological sciences affiliated with UCLA, University of Washington, and General Electric, which operated the Hanford Nuclear Reservation at this time, as well as regional specialists from the U.S. Fish and Wildlife Service and the U.S. Public Health Service/Arctic Public Health Service.
50. UCRL was renamed Lawrence Radiation Laboratory, Livermore (while still affiliated with the University of California) following Ernest Lawrence's death in 1958.
51. *Proceedings of the Second Plowshare Symposium, May 13–15, 1959*, San Francisco, Calif., Part II: Excavation, UCRL-5676, DOE doc. NV0015962, NTA. Quotation is from the preface. No proceedings are available from the symposium's first day, suggesting that this portion of the conference was probably classified and dealt with the phenomenology of the "peaceful nuclear explosives" themselves. It was normal practice for the national laboratories to hold such classified/unclassified conferences; typically, as Peter Westwick notes, conference attendance was higher for the classified days. Peter J. Westwick, "Secret Science: A Classified Community in the National Laboratories," *Minerva* 38 (2000): 363–391.
52. *Proceedings*.
53. Cratering data was not collected for these tests because there was little interest in these processes at the time and because erosive wave action quickly modified crater dimensions.
54. Data came primarily from two 1.2-kt shots in alluvial fill, code-named Buster-Jangle and Teapot Ess, and one very small 90-ton explosive Neptune in tuff. O'Neill (*Firecracker Boys*) may have overemphasized the significance of the October 1958 Neptune test in the downward revisions of the Chariot design. Although the Neptune test indeed vented a much smaller percentage of its radioactivity into the atmosphere than the other cratering shots, suggesting to Plowshare scientists that, at an optimum depth of burial, radioactivity could be substantially reduced while maintaining a relatively large crater size, this experiment was not vital or decisive in turning Chariot into a 100-kt-range experiment from the original megaton-range harbor project. These revisions to the harbor plan began, in effect, soon after Teller and Johnson returned from Alaska in July 1958. In addition, data from Neptune were discounted as potentially insignificant by at least one Livermore scientist for a number of reasons. M. A. Harrison, "Excavation with Nuclear Explosives," in *Proceedings*, 20–28.

55. Harrison, "Excavation," 25.
56. Ibid.
57. Ibid., 21.
58. Ibid. For a kiloton nuclear project, cost estimates were at $5 per yard of earth moved; for a megaton project, this cost decreased to one cent per yard. Chemical explosives at the same time cost roughly fifty cents per yard for projects of any size.
59. Clifford Bacigalupi, "Excavation of a Harbor (Project Chariot)," in *Proceedings*, 67–70.
60. A. Vay Shelton, "Radioactivity and Earth Motion," in *Proceedings*, 29–33. Shelton's calculations had been based on the four Nevada tests mentioned above.
61. Harrison, "Excavation with Nuclear Explosives," 28.
62. Bacigalupi, "Excavation of a Harbor ," 67, emphasis added.
63. Ibid.
64. On the constructed and contextual nature of maps, and on questions of subjective selection versus mimetic representation in the history of cartography, see J. B. Harley, *New Nature of Maps: Essays in the History of Cartography* (Baltimore: Johns Hopkins University Press, 2002); John Pickles, *A History of Spaces: Cartographic Reason, Mapping, and the Geo-coded World* (London: Routledge, 2004).
65. Bacigalupi, "Excavation of a Harbor ," 67.
66. John N. Wolfe, "The Ecological Aspects of Project Chariot," in *Proceedings*, 60–66. Quotation from p. 60.
67. Ibid., 60. Emphasis in original. The human economy in the Arctic, as we will examine in the next chapter (and as Wolfe recognized at least in his Arctic system diagram), depended on the land as well as on the sea.
68. Ibid., 60–61, 62.
69. Ibid., 65, 64.
70. Teller's speech was delivered at the University of Alaska Fairbanks commencement, in which he was presented with an honorary doctorate from President Ernest Patty, who was evidently eager to pursue a working friendship between the AEC and his university. Cited in O'Neill, *Firecracker Boys*, 89.
71. Plowshare Program, *Project Chariot:* An outline of Phase-IV (1961 season) Activities, July 21, 1961, Special Projects Division, San Francisco Operations Office, U.S. Atomic Energy Commission, DOE doc. NV0092199, NTA; O'Neill, *Firecracker Boys*.
72. U.S. AEC meeting minutes, May 22, 1959, DOE doc. NV0000753, NTA.
73. U.S. AEC, San Francisco Operations Office, "Statement of Committee on Environmental Studies for Project Chariot," January 7, 1960, Don C. Foote Collection, Box 11, Folder 23, Archives, University of Alaska Fairbanks.
74. Ibid.
75. Don Foote, handwritten notes in margins of U.S. AEC, "Statement of Committee on Environmental Studies for Project Chariot," January 7, 1960, Don C. Foote Collection, Box 11, no. 23, University of Alaska Fairbanks; Johnson, Pruitt, Swartz, and Viereck to Committee on Environmental Studies, Project Chariot, June 8, 1960, Project Chariot Collection, Box 1, Folder 1, Archives, University of Alaska Fairbanks. These contradictions would be corroborated in subsequent articles by Don Foote, Bill Pruitt, Les Viereck, and Barry Commoner in a special issue of *Nuclear Information* 3 (June 1961). The conflicting representations of the Cape Thompson environment will be discussed more extensively in the next chapter.
76. Bacigalupi, "Excavation of a Harbor."
77. Wolfe to Kessel, March 20, 1961, Project Chariot Collection, Box 1, Folder 1, Archives, University of Alaska Fairbanks. Emphasis in original.

78. Viereck to Wood, December 29, 1960, Project Chariot Collection, Box 1, Folder 6, Archives, University of Alaska Fairbanks.
79. Peter Coates, "Project Chariot: Alaskan Roots of Environmentalism," *Alaska History* (1989): 1–31.
80. Teller to Fidler, October 7, 1958.
81. U.S. AEC press release, June 9, 1958.

3. *Geographies of Authority*

1. Anderson to Foote, March 19, 1959, Don C. Foote Collection, Box 19, Correspondence: Alaska Rural Development Board, 1958–1959, University of Alaska Fairbanks; Dan O'Neill, *The Firecracker Boys* (New York: St. Martin's Press, 1994), 262–264. See also Scott Kirsch and Don Mitchell, "Earth-moving as the 'Measure of Man': Edward Teller, Geographical Engineering, and the Matter of Progress," *Social Text* 54 (1998): 100–134; "Don Charles Foote," n.d., [1968], Don C. Foote Collection, Box 31, curriculum vitae, University of Alaska Fairbanks.
2. Foote to Johnson, March 24, 1959, Don C. Foote Collection, Box 25, Correspondence: Atomic Energy Commission, 1959, Archives, University of Alaska Fairbanks.
3. Wolfe to Foote, May 20, 1959; Wolfe to Foote, May 5, 1959, Don C. Foote Collection, Box 25, Correspondence: Atomic Energy Commission, 1959, Archives, University of Alaska Fairbanks.
4. Foote to Division of Biology and Medicine, AEC, June 8, 1959, Don C. Foote Collection, Box 25, Correspondence: Atomic Energy Commission, 1959, Archives, University of Alaska Fairbanks; Foote to Division of Biology and Medicine, AEC, May 26, 1959, Don C. Foote Collection, Box 25, Correspondence: Atomic Energy Commission, 1959, Archives, University of Alaska Fairbanks. Foote's proposal also included objectives for a second phase of research from May 1960 to May 1961.
5. Foote planned to return to McGill following his stint in Point Hope, and to use data collected in the research as the basis for a doctoral dissertation. Foote to Hare, January 10, 1960, Don C. Foote Collection, Box 21, Correspondence: Hare, Kenneth, Archives, University of Alaska Fairbanks.
6. Foote to Campbell, October 23, 1959, Don C. Foote Collection, Box 25, Correspondence: Atomic Energy Commission, 1959, Archives, University of Alaska Fairbanks.
7. Don Foote, "The Economic Base and Seasonal Activities of Some Northwest Alaskan Village: A Preliminary Report," December 1, 1959, U.S. Atomic Energy Commission PNE-440, Oak Ridge; Don Foote, "The Eskimo Hunter at Point Hope, Alaska, Part I: September, 1959, to May, 1960" and "The Eskimo Hunter at Point Hope, Alaska, Part II: May to September, 1960," U.S. Atomic Energy Commission, DOE doc. NV0173276, NTA; Don Foote and Berit Arnstad Foote, field notes, Don C. Foote Collection, Box 12, AEC Project Chariot: Sea-Ice Project, University of Alaska Fairbanks; Kirsch and Mitchell, "Earth-moving."
8. Foote to Campbell, November 26, 1959, cited in O'Neill, *Firecracker Boys*, 102.
9. Foote, "The Eskimo Hunter at Point Hope, Alaska, Part I"; "over 10,000 miles" from Foote to Miller, November 17, 1960, Don C. Foote Collection, Box 19, Correspondence: Committee for Nuclear Information, Archives, University of Alaska Fairbanks.
10. Foote, "The Eskimo Hunter at Point Hope, Alaska, Part I." Some cash also came into Point Hope from wage labor in larger Alaska towns and cities during the summer, and from meat sales.
11. Ibid. Emphasis in original.
12. Foote, "Economic Base and Seasonal Activities." Although this report was sent to the committee by Don's brother, Joe Foote, it should have been received by the

committee no later than December 7, 1959. Foote to Campbell, November 26, 1959, Don C. Foote Collection, Box 25, Correspondence: Atomic Energy Commission, 1959, Archives, University of Alaska Fairbanks. Don sent his reports to Joe, then in Providence, Rhode Island, for editing.

13. Foote, "Economic Base and Seasonal Activities." Foote was referring specifically, in this case, to the village of Kivalina, but he expressed similar views on the significance of hunting activities to the Eskimo way of life in regards to Point Hope and other villages in Foote, "Economic Base and Seasonal Activities" and "The Eskimo Hunter at Point Hope, Alaska," parts 1 and 2.
14. Foote to Osgood, January 20, 1960, University of Alaska Fairbanks. Don C. Foote Collection, Box 21, Correspondence: Rev. Thomas Osgood, Archives, University of Alaska Fairbanks. For a related point, see Foote to Campbell, November 26, 1959.
15. Foote to Campbell, November 26, 1959.
16. O'Neill, *Firecracker Boys*, 105–106.
17. Campbell to Foote, November 19, 1959, Don C. Foote Collection, Box 25, Correspondence: Atomic Energy Commission, 1959, Archives, University of Alaska Fairbanks.
18. Foote to Campbell, November 26, 1959.
19. T. E. Frankson, "Meeting Minutes of the Native Village of Point Hope, Alaska, November 30, 1959," DOE doc. NV0316527, NTA. English was a second language for Point Hope Eskimos (who spoke Inupiaq), but was spoken more commonly by younger than by older generations.
20. Frankson, "Meeting Minutes."
21. Foote to Campbell, November 29, 1959, Don C. Foote Collection, Box 25, Correspondence: Atomic Energy Commission, 1959, Archives, University of Alaska Fairbanks.
22. U.S. AEC press release, Chariot 1–60, March 4, 1960, DOE doc. NV0316557, NTA.
23. Weaver to Office of Test Operations Folders, "Report on Trip to Alaska during Period March 2 to March 11, 1960," March 17, 1960, DOE doc. NV0078632, NTA.
24. O'Neill, *Firecracker Boys*, 112–130. Although O'Neill was referring specifically to the meeting in Point Hope, the observation suggests the likelihood of communications problems at the other villages as well, where, without Don Foote on hand, the people would likely have been less informed about Chariot to begin with than those at Point Hope.
25. King to Foote, April 3, 1960, Don C. Foote Collection, Box 21, Folder 82, Archives, University of Alaska Fairbanks.
26. "Excerpts of a tape recording made by Rev. Keith Lawton of the Episcopal Mission in Point Hope, Alaska, during public information meeting given by members of the USAEC and the Environmental Committee for Project Chariot, on March 14, 1960," Don C. Foote Collection, Box 12, Folder 29, Archives, University of Alaska Fairbanks. O'Neill provides a more detailed account of this meeting, quoting extensively from the tape recording, in his chapter "The AEC Meets the Eskimos," *Firecracker Boys*, 112–130. Largely because the Inupiaq Eskimo language of the region had no written form, O'Neill notes, reel-to-reel tape recorders had been widely adopted by the Eskimos as a means of sending oral "letters" to friends and family living in other villages.
27. Reverend Lawton played a key role in bringing information on nuclear testing and fallout to Point Hope, and also later in publicizing Point Hope's plight in his native New England.
28. Don Foote to Joe Foote, June 21, 1961, Don C. Foote Collection, Box 20, Correspondence: Foote, Joe: 15 June 1961–March 1965, Archives, University of Alaska Fairbanks.

29. Tape recording made by Lawton of March 14, 1960, meeting at Point Hope.
30. O'Neill, *Firecracker Boys*, 147. Italics are O'Neill's, based on the audio tape of the meeting.
31. Tape recording made by Lawton of March 14, 1960, meeting at Point Hope.
32. According to Les Viereck, who heard the audio tape of the meeting. Viereck to Francis, July 1, 1961, William O. Pruitt Collection, Box 1, Folder 2, Archives, University of Alaska Fairbanks.
33. U.S. AEC press release, Chariot 2–60, March 4, 1960, Don C. Foote Collection, Box 11, Folder 23, Archives, University of Alaska Fairbanks. The Plowshare Advisory Committee was established in 1959. Its members were "drawn from scientific, Government and business fields [to] advise the Commission on the general orientation and policies of the Plowshare program and specific aspects of work on experimental programs." U.S. AEC press release, November 12, 1959, DOE doc. NV0050885, NTA. The January 7 statement is quoted in chapter 2.
34. Weaver, "Report on Trip"; Foote to Cooke, March 16, 1960, Don C. Foote Collection, Box 20, Correspondence: Cooke, Alan, Archives, University of Alaska Fairbanks; Foote to A. Johnson, March 15, 1961, Don C. Foote Collection, Box 21, Folder 76, Archives, University of Alaska Fairbanks; Johnson, Pruitt, Swartz, and Viereck to Committee on Environmental Studies, Project Chariot, June 8, 1960, Project Chariot Collection, Box 1, Folder 1, Archives, University of Alaska Fairbanks.
35. Foote to Cooke, March 16, 1960.
36. Weaver, "Report on Trip."
37. "A.E.C. Is Pursuing Harbor Project," *New York Times*, March 13, 1960.
38. Johnson, Pruitt, Swartz, and Viereck to Committee, June 8, 1960.
39. Ibid.
40. Foote to Committee on Environmental Studies for Project Chariot, June 17, 1960, Don C. Foote Collection, Box 25, Correspondence: Atomic Energy Commission, 1960, Archives, University of Alaska Fairbanks.
41. Foote to Wolfe, June 17, 1960, Don C. Foote Collection, Box 25, Correspondence: Atomic Energy Commission, 1960, Archives, University of Alaska Fairbanks.
42. Wolfe to Johnson, Pruitt, Swartz, and Viereck, July 11, 1960, Don C. Foote Collection, Box 24, Correspondence: Atomic Energy Commission, Project Chariot, Archives, University of Alaska Fairbanks. Emphasis in original.
43. Wolfe to Foote, June 23, 1960, Don C. Foote Collection, Box 25, Correspondence: Atomic Energy Commission, 1960, Archives, University of Alaska Fairbanks.
44. U.S. AEC, "Visit of News Media Representatives to Cape Thompson Area in Connection with Plowshare's Chariot Proposal," August 12, 1960, DOE doc. NV0108852, NTA.
45. Viereck to Foote, August 27, 1960, Don C. Foote Collection, Box 22, Folder 155, Archives, University of Alaska Fairbanks.
46. Lawrence E. Davies, "Proposed Atomic Blast in Arctic Is Called Safe," *New York Times*, August 17, 1960.
47. Ibid.
48. Viereck to Wood, December 29, 1960, Project Chariot Collection, Box 1, Folder 6, Archives, University of Alaska Fairbanks.
49. Ibid.
50. "Staff Comments on Letter of 29 December 1960 from Mr. Leslie A. Viereck to Dr. William R. Wood," n.d., Plowshare Box P-50, LLNL. No authorship is given for this report, but it was most likely composed by staff at the AEC San Francisco Operations Office.
51. Ibid.

52. Foote to Wolfe, March 16, 1961, Don C. Foote Collection, Box 25, Folder 183, Archives, University of Alaska Fairbanks.
53. See Susan Millar and Don Mitchell, "Spectacular Failure, Contested Success: The Project Chariot Bioenvironmental Program," *Ecumene* 5 (1998): 287–302; O'Neill, *Firecracker Boys*. For its part, the University of Alaska at Fairbanks is now the repository for extensive Project Chariot archival and oral history collections at its Rasmussen Library, and in 1993 it awarded honorary doctorates to Bill Pruitt and Les Viereck, the two biologists that it had fired from the university in connection with their work on Chariot.
53. Foote to Pruitt, undated letter [probably fall 1960]; Foote to Pruitt, June 15, 1960; Foote to Pruitt, September 22, 1960; Foote to Pruitt, December 5, 1960; see Pruitt to Foote, May 2, 1960; Pruitt to Foote October 28, 1960. Don C. Foote Collection, Box 22, Correspondence: Pruitt, William O,' Archives, University of Alaska Fairbanks.
54. Foote to Viereck, September 29, 1960, Don C. Foote Collection, Box 22, Folder 155, Archives, University of Alaska Fairbanks.
55. Foote to Viereck, September 22, 1960, Don C. Foote Collection, Box 22, Folder 155, Archives, University of Alaska Fairbanks.
56. Foote to Pruitt, September 22, 1960.
57. Foote to Viereck, October 9, 1960, Don C. Foote Collection, Box 22, Folder 155, Archives, University of Alaska Fairbanks. Emphases in original.
58. Don Foote, "Chariot Major Policy Shifts," typed timeline and handwritten notes, Don C. Foote Collection, Box 12, no. 31, University of Alaska Fairbanks; see also Don C. Foote Collection, Box 11, no. 26, Archives, University of Alaska Fairbanks.
59. Bolson to Foote, December 19, 1960, Don C. Foote Collection, Box 25, Correspondence: Atomic Energy Commission, 1960, Archives, University of Alaska Fairbanks.
60. Schank to Foote, n.d., Don C. Foote Collection, Box 25, Correspondence: Atomic Energy Commission, 1960, Archives, University of Alaska Fairbanks.
61. Foote quickly offered these observations to Alaska senator Bartlett and to the governor's office. Replying to Bartlett's subsequent inquiry to the AEC, the AEC general manager insisted that "Mr. Foote is the only member [of the bioenvironmental program] to develop this extreme attitude." Yet soon after Bartlett received this reply, the senator heard news of Viereck's letter of resignation, and of the June 1960 letter of protest by Johnson, Swartz, Pruitt, and Viereck. General Manager (AEC) to Bartlett, February 6, 1961; General Manager (AEC) to Bartlett, March 27, 1961, Don C. Foote Collection, Box 19, Correspondence: Bartlett, E.L., Archives, University of Alaska Fairbanks; Foote to Riley, March 17, 1961, Don C. Foote Collection, Box 19: Correspondence, Alaska, Office of the Governor, Archives, University of Alaska Fairbanks.
62. Foote to Lachenbruch, November 25, 1960, Don C. Foote Collection, Box 21, Correspondence: Lachenbruch, Art, Archives, University of Alaska Fairbanks. Emphasis in original.
63. Ibid.
64. Pruitt to Foote, December 1, 1960, Don C. Foote Collection, Box 22, Correspondence: Pruitt, William O., Archives, University of Alaska Fairbanks.
65. Foote to Pruitt, December 5, 1960; November 17, 1960. Don C. Foote Collection, Box 22, Correspondence: Pruitt, William O., Archives, University of Alaska Fairbanks. Foote's suspicions of government surveillance were aroused by the presence of a dubious freelance writer and native Alaskan art enthusiast who had also been asking about Foote around Dartmouth's northern studies program, Foote's undergraduate program with which he maintained close ties and from which he hired research assistants. In addition, Foote believed his mail had been tampered with.

D. Foote to J. Foote, February 9, 1961, Don C. Foote Collection, Box 20, Correspondence: Foote, Joe, 1959–June 1961, Archives, University of Alaska Fairbanks; J. Foote to D. Foote, February 15, 1961, Don C. Foote Collection, Box 12, no. 36, Archives, University of Alaska Fairbanks; Foote to Cooke, February 9, 1961; Cooke to Foote, February 13, 1961, Don C. Foote Collection, Box 20, Correspondence: Cooke, Alan, Archives, University of Alaska Fairbanks. According to Dan O'Neill (*Firecracker Boys*, 202–205), while the FBI claims not to maintain files on Foote, the CIA did maintain a file but denied freedom of information requests to declassify it.

66. Minutes from AEC meeting 1697, February 8, 1961, DOE doc. NV0017855, NTA. See also Atomic Energy Commission meeting 1742, May 31, 1961, DOE doc. NV0000768, NTA.
67. Minutes from AEC meeting 1697, February 8, 1961.
68. U.S. AEC, Summary Notes of Meeting of the Commissioners with the Plowshare Advisory Committee, February 15, 1961, DOE doc. NV0075510, NTA. At Plowshare Advisory Committee meetings, panel members were typically outnumbered by a two- or three-to-one ratio by personnel from Livermore, the Sandia engineering laboratory, AEC San Francisco Operations Office, and other constituents of the classified Plowshare community. "Third Plowshare Advisory Committee Meeting, May 25 and 26, 1960," June 1, 1960, DOE doc. NV0134557, NTA; "Presentations on Public Safety to the Meeting of the Plowshare Advisory Committee, October 19–20, Sandia Corp., Albuquerque, NM," DOE doc. NV0016347, NTA.
69. Cited in O'Neill, *Firecracker Boys*, 190.
70. For more on Kessel's role in administering the University of Alaska contract, and on her related conflicts with Bill Pruitt over what he considered the censorship of his work, see Millar and Mitchell, "Spectacular Failure, Contested Success."
71. Pruitt had actually written to committee member Kermit Larson as early as March 1959, as the bioenvironmental program was initially developed, to suggest "a complete study of lichen physiology in your proposed studies at Cape Thompson and in the laboratory." Pruitt to Larsen, March 25, 1959, William O. Pruitt Collection, Box 2, Folder 9, Archives, University of Alaska Fairbanks.
72. U.S. Atomic Energy Commission, *Bioenvironmental Features of the Ogotoruk Creek Area, Cape Thompson, Alaska: A First Summary by the Committee on Environmental Studies for Project Chariot*, December 1960; cited in O'Neill, *Firecracker Boys*, 190. Emphasis added.
73. Interview with Barry Commoner, July 29, 2004.
74. As even John Wolfe would later describe CNI, "According to all reports and knowledge available to us, this organization is a serious-minded group, showing a genuine and serious concern about problems generated in the nuclear age." John Wolfe, "Comments on Biological Aspects of Report by St. Louis Citizens' Committee for Nuclear Information," as attached to Atomic Energy Commission, "St Louis Committee for Nuclear Information Bulletin on Project Chariot, Note by the Acting Secretary," August 15, 1961, AEC 811/82, DOE doc. NV0075525, NTA.
75. *Nuclear Information* 3 (4–7), June, 1961.
76. Michael W. Friedlander, "Predictions of Fallout from Project Chariot," *Nuclear Information* 3 (4–7), June 1961, 5–8.
77. About 10–22 mc/mi^2 (millicuries per square mile) as compared with an average of roughly 50 mc/mi^2 in the rest of the United States; Barry Commoner, "Biological Risks from Project Chariot," *Nuclear Information* 3 (4–7), June 1961, 9–12.
78. Ibid.
79. "Caribou May Bar Alaska A-Blasts," *New York Times*, June 4, 1961.
80. U.S. AEC, *Bioenvironmental Features;* cited in "A.E.C. Backs Plan for Alaska Blast," *New York Times*, June 9, 1961.

81. "A.E.C. Backs Plan for Alaska Blast."
82. H. Margolis, "Project Chariot: Two Groups of Scientists Issue 'Objective' but Conflicting Reports," *Science* 133: (June 23, 1961) 2000–2001.
83. Evidence suggests that some of this critique had been supplied to *Science* editor Howard Margolis in advance, or at least that Margolis had read AEC criticisms of the CNI report before reading the report itself. O'Neill, *Firecracker Boys*, 214–215.
84. Barry Commoner, M. W. Friedlander, and Eric Reiss, "Project Chariot" letter to the editor, *Science* 134 (August 18, 1961): 495–500. See also the reply by Howard Margolis, pp. 500–503.
85. AEC, "St Louis Committee for Nuclear Information Bulletin on Project Chariot."
86. Wolfe, "Comments on Biological Aspects."
87. Ibid.
88. Ulrich Beck, *World Risk Society* (London: Polity, 1999), 60.
89. AEC, "St Louis Committee for Nuclear Information Bulletin on Project Chariot."
90. Wolfe, "Comments on Biological Aspects."
91. Permissible dose limits were revised downward repeatedly during this period (including 1951, 1955, 1957, 1960, 1961, and 1964) by the AEC's advisory committee to the Division of Biology and Medicine and later by the Federal Radiation Council in response to new knowledge as well as new testing practices. AEC, "Some Thoughts on Radiation Protection Criteria at the Nevada Test Site," n.d., DOE doc. NV0068015, NTA. In September 1961, the Federal Radiation Council recommended to President Kennedy that the public's intake of "internal emitters" of radiation—from fission products such as strontium-90, strontium-89, and iodine-131—be reduced to one-third of the level previously set for individuals. "Council Proposes a 'Radiation Guide,'" *New York Times*, September 22, 1961.
92. Johnson to Philip, March 31, 1961, DOE doc. NV0005855, NTA.
93. John Kelly, "Report on Trip of July 24–August 2 to Livermore, California and Alaska," August 7, 1961, DOE doc. NV0075527, NTA.
94. Commoner, "Biological Risks from Project Chariot."
95. Reeves to Philip, April 28, 1961, "Fallout Patterns—NTS and Chariot," DOE doc. NV0039867, NTA.
96. Kelly, "Report on Trip."
97. Ibid.
98. Ibid.
99. Health Council of Point Hope to John Kennedy, March 3, 1961. Included in Weber to Kennedy, March 12, 1961, DOE doc. NV016872, NTA.
100. Contact with native rights attorney and activist Laverne Madigan and the Association on American Indian Affairs was established thanks largely to the work of a group of anti-Chariot New England activists who had heard about, and taken up, the cause after meeting the Lawtons and Point Hope Inupiat Antonio Weber in New Hampshire in the summer of 1960. See O'Neill, *Firecracker Boys*, 206–226.
101. Francis to Viereck, August 14, 1961, William O. Pruitt Collection, Box 1, Folder 2, Archives, UAF, emphasis in original. Sharon Francis, the confidential secretary to Secretary of the Interior Stewart Udall, became the key federal government contact person for the Chariot opposition in Alaska.
102. Members of the Village Council, Point Hope, Alaska, to Udall, July 25, 1961. Cited in O'Neill, *Firecracker Boys*, 230.
103. Francis to Viereck, August 14, 1961.
104. U.S. AEC, "Progress Report on Selected Programs to the Joint Committee on Atomic Energy," DOE doc. NV0075760, NTA; U.S. AEC, "Annual Report to Congress of the Atomic Energy Commission for 1961," January 1962, DOE doc. NV0032523, NTA. The scrubbing of technicians following the Gnome venting

was not mentioned in the report to Congress (at least not in the publicly accessible deleted version), but it was mentioned to me by one of the former AEC technicians, by this time long retired, who was a fellow traveler during a public tour of the Nevada Test Site in June 1996.

105. Bill Becker, "U.S. A-Bomb Test for Peaceful Use Frees Radiation," *New York Times*, December 11, 1961. Interestingly, the *Times* had planned a promotional "man in the news" feature on the AEC's Division of Peaceful Nuclear Explosives director John Kelly to accompany the Gnome report, which it decided to publish anyway. "Peaceful-Atom Physicist," *New York Times*, December 11, 1961.
106. U.S. AEC, "Progress Report on Selected Programs to the Joint Committee on Atomic Energy." See also U.S. AEC, "Annual Report to Congress of the Atomic Energy Commission for 1961."
107. The AEC's Ernest Campbell, the committee secretary, perhaps missing the forest for the trees, was particularly upset that Joseph Foote seemed to have had access to brother Don's AEC-contracted final report before the AEC itself did. In Philip to Kelly, "Commentary on *Harper's* Chariot Article," April 27, 1962, Plowshare Box P-50, LLNL; see also Kelly to Symington, May 3, 1962, Plowshare Box P-50, LLNL.
108. Paul Brooks and Joseph Foote, "The Disturbing Story of Project Chariot," *Harper's Magazine*, April 1962, 60–67, quotation from p. 67. Emphases in original.
109. Cf. Graves to President, University of California, April 19, 1962, Plowshare Box P-50, LLNL; Kelly to Symington, May 3, 1962.
110. Kelly to Batzel, March 16, 1962, Plowshare Box P-50, LLNL.
111. Foster to Kelly, April 30, 1962, Plowshare Box P-50, LLNL.
112. Kelly to Luedecke, May 23, 1962. DOE doc. NV0176830, NTA.
113. John Kelly, "Atomic Energy Commission, Plowshare Program (Project Chariot): Report to the General Manager by the Director of Peaceful Nuclear Explosives," July 10, 1962, Plowshare Box P-50, LLNL.
114. Foster to Kelly, April 27, 1962, Plowshare Box P-50, LLNL.
115. Foster to Kelly, April 27, 1962. The final April 30 version did not contain the "perhaps never stated" phrase, but instead included as one of six project goals for Chariot, "to dramatically demonstrate the capability and safety of the constructive earth moving power of nuclear explosives." Foster to Kelly, April 30, 1962, Plowshare Box P-50, LLNL
116. Foster to Kelly, April 30, 1962.
117. Kelly, "Atomic Energy Commission, Plowshare Program (Project Chariot)."
118. Don Foote to Joe Foote, March 1, 1961, Don C. Foote Collection, Box 20, Correspondence: Foote, Joe, 1959–June 1961, Archives, University of Alaska Fairbanks. Emphasis in original.
119. The native land rights movement in Alaska led to the Alaska Native Claims Settlement Act of 1971. Don Foote, for his part, returned to McGill, where he completed his Ph.D. in 1965, and worked as an assistant professor in the Geography Department until 1968, when he gained a position at the Institute for Social Economic and Government Research, against the wishes of university president William Wood at the University of Alaska. Foote continued to work extensively in the field in polar regions during this time, including reports for Canadian government planning and resource agencies such as the Fisheries Board of Canada and the Department of Indian Affairs and Northern Development. Foote also continued his involvement in scientific/academic politics. In February 1969, just days before a "research halt" was to take place that he had organized to protest the University of Alaska's development plans in connection with the discovery of North Slope oil, Don Foote died after an auto accident in Fairbanks at the age of thirty-seven. See O'Neill, *Firecracker Boys*, 262–264. Although no clear evidence has been produced

to indicate a conspiracy, Foote's death, according to one contemporary researcher, is still commonly remembered by the Inupiat of Point Hope as an assassination.

4. Nuclear Craters

1. U.S. AEC, "Project Managers Report: Project Sedan," May 1963, AEC-LIV-59, Plowshare Box P-5, LLNL; Milo Nordyke and M. M. Williamson, "The Sedan Event: Final Report," August 6, 1965, PNE-242, Plowshare Box P-6, LLNL; Southwestern Radiological Health Laboratory, "Iodine Inhalation Study for Project Sedan," Public Health Service, received at Lawrence Radiation Laboratory, Livermore, August 3, 1964, Plowshare Box P-6, LLNL.
2. U.S. AEC, "Project Managers Report: Project Sedan." Although almost three times smaller than the final design for Chariot's explosive yield, Sedan still had roughly seven times the force of the Hiroshima explosion. The motion picture footage of the Sedan explosion can be viewed on the website of the Nevada office of the U.S. Department of Energy: http://www.nv.doe.gov/news%26pubs/photos%26films/0800030/default.htm.
3. Nordyke and Williamson, "The Sedan Event."
4. U.S. AEC press release, July 7, 1962, Plowshare Box P-5, LLNL.
5. "H-Blasted Crater Twelve Million Tons Deep Makes a Lab for AEC Men," *Life* 54 (June 21, 1963): 47–50, 50.
6. U.S. AEC press release, August 24, 1962, Plowshare Box P-5, LLNL.
7. Cf. "Man's First Atomic Cave," *Life* 54 (March 15, 1963): 55; "Digging with H-Bombs; Project Sedan," *Business Week*, May 18, 1963, 154; "Another Panama Canal? A-Blasts May Do the Job," *U.S. News and World Report*, June 10, 1963, 74–75.
8. The photo shoot from *Life* was part of a coordinated AEC public information plan for media visits to Sedan Crater on May 6 and 7, 1963. Duncan Clark, "News Media Visit to Sedan Crater," April 19, 1963, DOE doc. NV0178693, NTA; U.S. AEC, "Press releases: Sedan news briefing," April 24, 1963, DOE doc. NV0038454, NTA. For a related point on the rhetoric of histories of experimentation advanced in educational science texts, see Geoffrey Cantor, "The Rhetoric of Experiment" in *The Uses of Experiment*, ed. D. Gooding, T. Pinch and S. Schaffer (Cambridge: Cambridge University Press, 1989).
9. U.S. AEC press release, July 7, 1962. Similar wording appears in U.S. AEC, Monthly report to the General Advisory Committee to the AEC, July 1962, DOE doc. 73795; "Review of the Plowshare Program for Fiscal Years 1963 and 1964," DOE doc. NV0176767, NTA.
10. U.S. AEC, "Project Managers Report: Project Sedan"; Nordyke and Williamson, "The Sedan Event."
11. Vermillion to Miller, "Project Manager's Completion Report—Project Sedan," October 30, 1962, DOE doc. NV0016821, NTA.
12. Reeves to distribution, "Planning Directive (PL-3-6-62)—Sedan, Revision I," June 22, 1962, DOE doc. NV0124501, NTA; Reeves to distribution, "Planning Directive (PL-3-6-62) Sedan, Revision II," July 2, 1962, DOE doc. NV0124497, NTA; Clark to Luedecke, "Proposed Announcement on Project Sedan," June 29, 1962, DOE doc. 137563. Photographs of the Sedan explosion and crater were eventually distributed as an AEC press release (U.S. AEC press release, November 28, 1962, DOE doc. NV0041354, NTA).
13. Nordyke and Williamson, "The Sedan Event."
14. Ibid.; U.S. AEC, "Project Managers Report: Project Sedan"; U.S. AEC, "Background Notes on Plowshare," December 1962, DOE doc. NV0179114, NTA.
15. Nordyke and Williamson, "The Sedan Event."

16. U.S. AEC, "Project Managers Report: Project Sedan"; Philip Fradkin, *Fallout: An American Nuclear Tragedy* (Tucson: University of Arizona Press, 1989).
17. U.S. AEC press release, October 5, 1962, Plowshare Box P-5, LLNL. Not everyone was fortunate enough to be indoors. For instance, as one public health official reported seven years later, one anonymous "lady in Lund [Nevada] is dying (or may have passed away by now) of terminal lung cancer." The woman had "expressed her feelings to local residents that it may be due to her being in the Sedan cloud. She was out and running around in the cloud trying to [move] her livestock indoors to 'protect' them, and she feels she inhaled a lot of the cloud dust in the process." Wruble to Coleman, "Off-site residents' concerns," September 19, 1969, DOE doc. NV0163629, NTA.
18. Quoted in Fradkin, *Fallout*, 215. These reflections came in the context of a 1979 deposition for a lawsuit (*Timothy v. United States of America*) filed by a cancer victim living in an area where Pendleton had measured high levels of fallout.
19. R. C. Pendleton, R. D. Lloyd, and C. W. Mays, "Iodine-131 in Utah during July and August 1962," *Science*, August 16, 1963, 640–642.
20. U.S. AEC, "Project Managers Report: Project Sedan." See also Trevor Findlay, *Nuclear Dynamite: The Peaceful Nuclear Explosions Fiasco* (Sydney: Brassey's Australia, 1990); Fradkin, *Fallout;* Dan O'Neill, *The Firecracker Boys* (New York: St. Martin's, 1994).
21. Fradkin, *Fallout*, 214–217; Thompson to Pendleton, August 4, 1962, DOE doc. 32692, cited in Fradkin, *Fallout*, 216.
22. Fradkin, *Fallout;* Terry Tempest Williams, *Refuge* (New York: Pantheon, 1991).
23. Knapp to Dunham, June 27, 1963, DOE doc. NV0153306, NTA.
24. Dunning to Bloch, "Recommendations of the Ad Hoc Working Group on Radioiodine in the Environment," April 25, 1963, DOE doc. NV0043348, NTA.
25. Knapp to Dunham, June 27, 1963, DOE doc. NV0153306, NTA. Emphasis added.
26. Harold Knapp, "Iodine-131 in Fresh Milk and Human Thyroids Following a Single Deposition of Nuclear Test Fallout," June 1, 1963, Division of Biology and Medicine, AEC, Washington, D.C., DOE doc. NV0153314, NTA.
27. Harold Knapp, "Conclusions of a Report on Observed Relations between the Deposition Level of Fresh Fission Products from Nevada Tests and the Resulting Levels of I-131 in Fresh Milk," included in Dunham to Langham, Comar, Gofman, Nielsen, and Placak, June 7, 1963, DOE doc. NV0153320, NTA. For more on Knapp and the Utah radioiodine controversy, see Scott Kirsch, "Harold Knapp and the Geography of Normal Controversy: Radioiodine in the Historical Environment," *Landscapes of Exposure: Knowledge and Illness in Modern Environments*, ed. G. Mitman, M. Murphy, and C. Sellers *Osiris*, Vol. 19: 167–181.
28. Dunning's model for estimating I-131 uptake after a nuclear war was based on data from the 1957 Sellafield reactor fire in the UK.
29. Dunning to Woodruff, June 14, 1963, "Comments on 'Iodine-131 in Fresh Milk and Human Thyroids Following a Single Deposition of Nuclear Test Fallout' by Dr. Harold Knapp," included in Woodruff to Dunham, June 17, 1963, DOE doc. NV0153315, NTA.
30. Dunning to Woodruff, June 14, 1963, "Comments." See also Higgins to Dunham, March 30, 1963, "Preliminary Comments on the Knapp I-131 Study," DOE doc. NV0160850, NTA.
31. Knapp to Dunham, June 27, 1963, DOE doc. NV0153306, NTA.
32. "The most direct information available," as Knapp later described it, "is limited to some coincidental measurements of iodine-131 levels in milk from two dairy herds and to the open-field external [gamma]-dose rate reported by Public Health Service monitors in the vicinity of the area where the cows were grazing." These measurements were made at Alamo and Caliente, Nevada, following the July 14, 1962, Small Boy test.

H. A. Knapp, "Iodine-131 in Fresh Milk and Human Thyroids Following a Single Deposition of Nuclear Test Fall-Out," *Nature* 202 (May 9, 1964): 534–537, 535.

33. "Radiation Safety Plan for Field Operations Portion of the Iodine Inhalation Study to Be Conducted by the U.S. Public Health Service in Connection with the Sedan Event of the Plowshare Program," n.d. (preshot planning document), DOE doc. NV0003763, NTA; Southwestern Radiological Health Laboratory, "Iodine Inhalation Study."
34. On "social distancing" in knowledge and representational strategies in public health, see Michael Brown, "Ironies of Distance: An Ongoing Critique of the Geographies of AIDS," *Society and Space* 13 (1995): 159–183; Kirsch, "Harold Knapp."
35. Dunning to Woodruff, June 14, 1963, "Comments."
36. Harold Knapp, Comments on Dunning to Woodruff, June 14, 1963, DOE doc. NV0161119, NTA; see R.C. Pendleton to Harold Knapp, March 26, 1963. In *Fallout, Radiation Standards, and Countermeasures: Hearings before the Subcommittee on Research, Development, and Radiation of the Joint Committee on Atomic Energy, Congress of the United States, Eighty-Eighth Congress, August 20, 21, 22, and 27, 1963*, part 2 (Washington, D.C.: U.S. Government Printing Office, 1963), 1034–1037; R.C. Pendleton to Harold Knapp, May 6, 1963, DOE doc. NV0027353, NTA; Grant S. Winn to Harold Knapp, March 28, 1963; April 1, 1963. In *Fallout, Radiation Standards, and Countermeasures*, part 2, 1039–1046.
37. Harold Knapp, Comments on Dunning to Woodruff; Knapp, "Iodine-131 in Fresh Milk and Human Thyroids"; see discussion of the Knapp-Pendleton correspondence in Fradkin, *Fallout*. Both scientists believed the AEC was holding up publication of their papers—in *Science* (Pendleton) and *Nature* (Knapp). See also Barton Hacker, *Elements of Controversy: The Atomic Energy Commission and Radiation Safety in Nuclear Weapons Testing, 1947–1974* (Berkeley: University of California Press, 1994), 222–226.
38. Dunning to Woodruff, June 14, 1963, "Comments."
39. Knapp, "Iodine-131 in Fresh Milk," *Nature;* H. A. Knapp, "Iodine-131 in Fresh Milk and Human Thyroids Following a Single Deposition of Nuclear Test Fall-Out," TID-19266, Health and Safety (TID-4500, 24th ed.), (Washington, D.C.: Fallout Studies Branch, Division of Biology and Medicine, Atomic Energy Commission, June 1, 1963).
40. "Report by Dr. H. A. Knapp, AEC, on 'Iodine 131 in Fresh Milk and Human Thyroids Following a Single Deposition of Nuclear Test Fallout,'" in Part 2, *Fallout, Radiation Standards, and Countermeasures*, part 2, 915–1031; "Ad Hoc Committee Review of Report on 'Iodine 131 in Fresh Milk and Human Thyroids Following a Single Deposition of Nuclear Test Fallout,'" in *Fallout, Radiation Standards, and Countermeasures*, part 2, 1076–1078; "Letter Concerning Comments Relative to the Committee Review of the Knapp Report from Dr. Harold A. Knapp, to John T. Conway, Executive Director, JCAE, Dated September 9, 1963," in *Fallout, Radiation Standards, and Countermeasures*, part 2, 1078–1082.
41. Kelly to Batzel, August 31, 1962, Plowshare Box P-5, LLNL.
42. Mays to Hamburger, June 26, 1963, DOE doc. NV0015389, NTA.
43. Pendleton, Lloyd, and Mays, "Iodine-131 in Utah." Infants were at greater risk of iodine irradiation because of "(a) the small size of the child's thyroid which increases its concentration of I^{131}, (b) its presumed greater sensitivity to irradiation at this stage, and (c) the long post-irradiation life span during which delayed effects could appear." Charles W. Mays, "Iodine-131 in Utah from Nuclear Tests in Nevada," August 20, 1963. Prepared for the fallout hearings of the Congressional Committee on Atomic Energy, August 20–21, 1963, at Washington, D.C., DOE doc. NV0011720, NTA.
44. Knapp, "Conclusions."

45. Mays to Hamburger, June 26, 1963. Emphasis in original.
46. *Fallout, Radiation Standards, and Countermeasures: Hearings before the Subcommittee on Research, Development, and Radiation of the Joint Committee on Atomic Energy, Congress of the United States, Eighty-Eighth Congress, June 3, 4, and 6, 1963* (part 1) (Washington, D.C.: Government Printing Office, 1963); *Fallout, Radiation Standards, and Countermeasures*, part 2.
47. Statement of Dr. Charles Mays, University of Utah, *Fallout, Radiation Standards, and Countermeasures*, part 2, 536–563; Mays, "Iodine-131 in Utah from Nuclear Tests in Nevada."
48 "Excerpts from the Transcript of the President's News Conference," *Washington Post,* August 21, 1963; Howard Simons, "Kennedy Doubts Claim Infants Hurt by Fallout," *Washington Post,* August 21, 1963.
49. Dunning to Bloch, "Recommendations of the Ad Hoc Working Group on Radioiodine in the Environment." See also Harold Knapp's extensive recommendations, which were submitted to the AEC's Division of Biology and Medicine and made available to the Ad Hoc Committee. H. A. Knapp to Dunham, "Recommendations for Additional Measures Related to the Evaluation and Control of Radioiodine from Nevada Nuclear Test," March 12, 1963, DOE doc. NV0027355, NTA.
50. Seaborg to Harriman, n.d. (stamped June–August, 1963, Seaborg Collection), DOE doc. NV0178844, NTA.
51. Interview with Herbert York, August 4, 2004. York, Livermore's first director, was by this time chancellor at the University of California, San Diego, and an advisor to the Arms Control and Disarmament Agency and the secretary of state. See Findlay, *Nuclear Dynamite*, 40–60.
52. Cited in Rudy Abramson, *Spanning the Century: The Life of W. Averell Harriman, 1891–1986* (New York: William Morrow and Company, 1992), 596.
53. Ibid., 598.
54. UK, USA, and USSR, Treaty Banning Nuclear Weapon Tests in the Atmosphere, in Outer Space, and Underwater (LTBT).
55. The only concession to Plowshare, and perhaps to future Soviet excavation plans, was a change in amendment procedures to require a simple (instead of two-thirds) majority for adopting an amendment to accommodate such international scale PNE projects. Findlay, *Nuclear Dynamite*.
56. U.S. Congress, Senate Foreign Relations Committee, *Hearings on the Nuclear Test Ban Treaty,* August 12–15, 19–23, 26–27, 1963, testimony of Edward Teller (Washington, D.C.: U.S. Government Printing Office, 1963), 455.
57. Transcript of National Broadcasting Company, "Meet the Press," September 1, 1963, vol. 7, no. 34. Produced by Lawrence Spivak, guest Dr. Edward Teller (Washington, D.C.: Merkle Press). Fears of a Soviet "Plowshare gap" were expressed in Edward Teller (with Allen Brown), *The Legacy of Hiroshima* (Garden City, N.Y.: Doubleday and Company, 1962); Ralph Sanders, *Project Plowshare: The Development of Peaceful Uses of Nuclear Explosives* (Washington, D.C.: U.S. Government Printing Office, 1962).
58. *Hearings*, testimony of Norris Bradbury.
59. *Hearings*, testimony of Teller, 433.
60. Who wrote these marginal asides or when they were written is unknown. But among the files of the beleaguered Plowshare program, they have the ring of authenticity. Kavanaugh to Bradbury and Foster, TWX 449-H Priority, n.d. "Treaty Banning Nuclear Weapon Tests in the Atmosphere in outer space and underwater," Plowshare Box P-31, LLNL. Capitals in original.
61. U.S. AEC, "Nuclear Excavation and the Test Ban Treaty," included in Seaborg to Rusk, no date, DOE doc. NV0075602, NTA.

62. Glenn T. Seaborg with Eric Seaborg, *Adventures in the Atomic Age* (New York: Farrar, Straus and Giroux, 2001), 230.
63. AEC, "Plowshare Excavation Program," July 10, 1962, AEC 811/105. DOE doc. NV0075573, NTA.
64. Ibid.
65. Meetings with the Army Corps of Engineers and other affiliated agencies had been initiated as early as 1959: cf. "Attendees Panama Meetings" and "Outline," September 24 and 25, 1959, Plowshare Box P-24, LLNL; "Trans-Isthmian Canal," October 30, 1959, Plowshare Box P-24, LLNL; Teller to Potter, telex, December 14, 1959, Plowshare Box P-24, LLNL.
66. Wilkes to Foster, March 21, 1963, Plowshare Box P-29, LLNL.
67. U.S. AEC, "Plowshare Program: Report to the General Manager by the Director, Division of Peaceful Nuclear Explosives," April 19, 1963, AEC 811/111. DOE doc. NV0075564, NTA.
68. Foster to Kelly, April 2, 1963, included in AEC, "Plowshare Program: Report to the General Manager."
69. AEC, "Plowshare Program: Report to the General Manager."
70. John Kelly, Memorandum for Commissioner Haworth through General Manage, "Discussion Paper on Plowshare Device Development," May 3, 1963, DOE doc. NV0075563, NTA. Seventeen underground nuclear tests in Nevada between November 1962 and May 1970 are classified as Plowshare device development tests. U.S. Department of Energy, *United States Nuclear Tests: July 1945 through September 1992*, DOE/NV-209 (Rev. 14), December 1994, Nevada Operations Office.
71. Kelly, Memorandum for Commissioner Haworth.
72. Technical Director's Plan for Project Schooner, August 5, 1963, revised August 23, 1963, Plowshare Box P-5, LLNL.
73. U.S. AEC, "Project Schooner: Description of Detonation," March 21, 1963, DOE doc. NV0133796, NTA; Miller to distribution, "Request for Review Panel Endorsement of Project Schooner Shot Sector," Plowshare Box P-5, LLNL; Public Health Service, "Preliminary Site Surveys—Project Schooner," June 11, 1963, Plowshare Box P-5, LLNL.
74. Gladwin Hill, "Atomic Excavations Are Planned by the A.E.C.," *New York Times*, June 24, 1964; Reeves to distribution, July 17, 1963, Plowshare Box P-5, LLNL; "Technical Director's Plan for Project Schooner"; Short to Nordyke, November 7, 1962, "Field Investigations of Possible Schooner Site," Plowshare Box P-5, LLNL.
75. "Technical Director's Plan for Project Schooner."
76. Becker and Minkler to Higgins, March 23, 1963, Plowshare Box P-5, LLNL; Martin to Higgins, March 21, 1963, Plowshare Box P-5, LLNL.
77. "Project Schooner Public Information Plan," draft, July 23, 1963, Plowshare Box P-5, LLNL.
78. Vermillion to Reeves, June 12, 1963, "Schooner Site Selection Activity in Idaho," June 12, 1963, Plowshare Box P-5, LLNL.
79. Ibid.
80. On the relations between Western Shoshone territoriality (as well as that of other Native American tribes) and the nuclear landscapes of Nevada and other parts of the western United States, see Valerie Kuletz, *The Tainted Desert: Environmental and Social Ruin in the American West* (London: Routledge, 1997). As one recent scientific study argues, even current U.S. Department of Energy dose reconstructions of radioactivity taken up by Native Americans during this period fail to adequately account for the exposures received by these groups, because of missing exposure pathways (such as small game) and other incorrect assumptions about food pathways. E. Frohmberg, R. Goble, V. Sanchez, and D. Quigley, "The Assessment of

Radiation Exposures in Native American Communities from Nuclear Weapons Testing in Nevada," *Risk Analysis* 20 (2000): 101–111.

81. Reeves to Clark, Kelly, Southwick, and Higgins, August 20, 1963, Plowshare Box P-5, LLNL; Reeves to distribution, August 22, 1963, "Public Discussion Regarding Schooner Site," Plowshare Box P-5, LLNL.
82. AEC to Johnson, Reeves, and Shute, October 8, 1963, Plowshare Box P-5, LLNL; Kelly to Luedecke, October 30, 1963. In AEC, "Revised Plowshare Program for FY 1964 and FY 1965," November 8, 1963, AEC 811/115, DOE doc. NV0075585, NTA.
83. Seaborg, *Adventures*.
84. Johnson to Brown, December 4, 1963, DOE doc. NV0176771, NTA.
85. "Technical Director's Plan for Project Sulky," November 8, 1963, revised October 30, 1964, Plowshare Box P-7, LLNL.
86. Johnson to Brown, December 4, 1963, DOE doc. NV0176771, NTA.
87. L. J. Vortman to distribution, "Transcontinental Canal," August 26, 1964, Plowshare Box P-24, LLNL.
88. Tad Szulc, "U.S. Decides to Dig a New Canal at Sea Level in Latin America and Renegotiate Panama Pact," *New York Times*, December 19, 1964.
89. H. P. Metzger, *The Atomic Establishment* (New York: Simon and Schuster, 1972), cited in Findlay, *Nuclear Dynamite*, 67.
90. William Beecher, "Panama Canal Traffic Tops Forecasts, Spurs Plans for New Passage," *Wall Street Journal*, March 28, 1961.
91. Lyndon B. Johnson, "Remarks of the President on the Panama Canal," December 18, 1964, Plowshare Box P-24, LLNL; Panama Canal Commission, PCC S/9 A 14 PF/1946, cited in Stephen Frenkel, "A Hot Idea? Planning a Nuclear Canal in Panama," *Ecumene* 5 (1998): 303–309.
92. Findlay, *Nuclear Dynamite*, 67–70; Frenkel, "A hot idea?"
93. Johnson, "Remarks."
94. U.S. AEC, "Test Manager's Operational Concept Project Sulky (Revised)," November 30, 1963, Plowshare Box P-7, LLNL.
95. Ibid.
96. U.S. AEC press release, December 21, 1964, DOE doc. NV0144921, NTA.
97. Johnson to Kelly, January 22, 1965, included in AEC, "Palanquin Event," February 5, 1965, AEC 811/125, DOE doc. NV0075604, NTA; Johnson to Kelly, January 22, 1965, "Concept Letter for Palanquin," COPKF 65–2, and "Total Project Plan, Project Palanquin," February 5, 1965, revised March 3, 1965, both Plowshare Box P-38, LLNL; "Test Manager's Operational Concept, Project Palanquin," February 5, 1965.
98. Seaborg to Bundy, March 24, 1965, DOE doc. NV0078203, NTA.
99. Kelly and Jacques to Vermillion, April 1, 1965, "Field Information Plan—Project Palanquin," Plowshare Box P-38, LLNL; LRL and NVOO, "Total Project Plan for Project Palanquin," February 5, 1965, revised March 3, 1965, Plowshare Box P-38, LLNL; Oliver de Lalla, "Palanquin Event: Quantitative Postshot Distribution of Close-In Fallout, Throwout, and Base Surge," August 5, 1969, UCRL-50631, Plowshare Box P-38, LLNL; NOAA, Weather Service Nuclear Support Office, Las Vegas, Nevada, "Project Palanquin: Weather Predictions and Surface Radiation Estimates for Project Palanquin," issuance date May 1989, PNE-912F, Plowshare Box P-38, LLNL.
100. Kelly and Jacques, "Field Information Plan."
101. Quotation from Francis Buck, "A Graphical Compilation of Data Collected Relative to the Palanquin Event, Rough Draft," SWRHL-25r, Southwestern Radiological Health Laboratory, U.S. Public Health Service, Plowshare Box P-38, LLNL; Glenn Seaborg, Memorandum for the President, "Preliminary Report on Project Palanquin," April 16, 1965, DOE doc. NV0031640, NTA.

102. Seaborg, "Preliminary Report on Project Palanquin"; Seaborg, "Follow-up to Memorandum of April 16 on Palanquin event, April 19, 1965," DOE doc. NV0031642, NTA.
103. Clark, Memo for Seaborg et al., April 26, 1965, "Questions and Answers on Palanquin," DOE doc. NV0075630, NTA; Labowitz to Crowes, May 5, 1965, DOE doc. NV0075626, NTA; Southwestern Radiological Health Laboratory, Memo: "Announced Event," April 19, 1965, DOE doc. NV0171137, NTA.
104. U.S. AEC, "Preliminary Report: Whole-Body Counting of Several Families in Nevada in Connection with the Palanquin Event of April 14, 1965," DOE doc. NV0038331, NTA. Soon after, the AEC "County Extension Agent" at Winnemucca, Nevada, and an associate of the "Nevada USDA State Defense Board" advised the AEC that: "there are rumors in the community about the dumping of milk, hazards to children from drinking of milk and the flying of some residents to the Nevada Test Site for whole-body counter." Public information countered these "rumors" by stressing that levels of radioiodine in milk were not necessarily a hazard, and that "no person did receive more than a permissible radiation exposure." Seaborg to Kastenmeier, May 8, 1965, DOE doc. NV0176644, NTA.
105. NOAA, "Project Palanquin."
106. "Thyroid Burdens in a Small Sample of the Off-Site Population following the Palanquin Event," n.d., DOE doc. NV0017176, NTA.
107. Teller, *Legacy of Hiroshima,* 64. Teller's remark was in reference to several experimental failures of the Lawrence Radiation Laboratory during the early 1950s.
108. Glenn Werth, "Schooner Planning Concept," May 12, 1965, Plowshare Box P-5, LLNL.

5. *Pragmatic Engineering Worlds*

1. Edward Teller, "University Role in Nuclear Explosives Engineering Research," in *Education for Peaceful Uses of Nuclear Explosives*, ed. Lynn E. Weaver (Tucson: University of Arizona Press, 1970), 294.
2. Edward Teller, Wilson K. Talley, Gary Higgins, and Gerald Johnson, *The Constructive Uses of Nuclear Explosives* (New York: McGraw-Hill Book Company, 1968), vi.
3. Livermore scientists and others affiliated with Plowshare presented (unclassified) scientific papers at numerous conferences during the late 1960s, including, for example, the Chicago Association of Commerce and Industry conference on Atomic Energy and Industry, Chicago, April 1968; Plowshare Symposium—Public Health Aspects, Las Vegas, October, 1968; Education for Peaceful Uses of Nuclear Explosives, Tucson, March–April 1969; Public Health Aspects of Peaceful Nuclear Explosives, Las Vegas, April 1969; American Nuclear Society, Engineering with Nuclear Explosives, Las Vegas, January 1970.
4. A 1968 survey showed that courses in "nuclear civil engineering" and engineering with "peaceful nuclear explosives" were in the curricula in at least twenty-two universities, and were incorporated as components in existing nuclear engineering courses in many more. Paul Kruger, "A Survey of University Courses in Nuclear Civil Engineering," in *Education for Peaceful Uses,* ed. Weaver 239–256.
5. Statement by John Kelly before the JCAE, January 5, 1965, Plowshare Box P-29, LLNL.
6. Milo Nordyke, "Discussion of the Need for a Harbor Demonstration," UOPKC 68–106, November 18, 1968, Plowshare Box P-22, LLNL.
7. AEC, "Report on the Tenth Meeting of the Plowshare Advisory Committee Held in Washington, D. C. on April 3 and 4, 1963," DOE doc. NV0064455, NTA.

8. Ernest Graves, Corps of Engineers, "Nuclear Excavation Research," n.d., Plowshare Box P-29, LLNL. This paper was probably written in 1963.
9. U.S. Army Engineer Nuclear Cratering Group, Corps of Engineers, "Engineering Manual on the Use of Nuclear Explosives for Large Scale Excavation," June 1963, revised draft, Plowshare Box P-29, LLNL.
10. Ibid.
11. Tennessee-Tombigbee Waterway Development Authority, "The Connecting Link: An Important Asset to the Expanding Industrial South" (brochure), City Hall Building, Columbus, Mo., n.d.), Plowshare Box P-7, LLNL. The Development Authority actually dated the lineage of the waterway idea to French settlers in the early 1800s. Corps feasibility studies, with unfavorable conclusions, were carried out in 1913, 1923, and 1932.
12. Suttle to Eastland, June 20, 1960, Plowshare Box P-7, LLNL; Suttle to Teller, June 20, 1960, Plowshare Box P-7, LLNL.
13. "Nuclear Blasts Are Proposed to Cut Tombigbee Waterway," *Clarion Ledger* (Nashville, Tenn.), September 26, 1961.
14. Wilkins to Seaborg, February 17, 1964, Plowshare Box P-7, LLNL.
15. Johnson to McNamara, March 5, 1964, Plowshare Box P-7, LLNL.
16. U.S. DOE, "Environmental Restoration. Offsites: Salmon Site, Mississippi," January 2000, Washington, D.C.: DOE
17. Johnson to Seaborg, March 5, 1964, Plowshare Box P-7, LLNL.
18. Johnson to Gibson, January 28, 1965, Plowshare Box P-7, LLNL.
19. Merritt to Miller, March 29, 1965, Plowshare Box P-7, LLNL.
20. Toman to Johnson and Higgins, January 15, 1965, Plowshare Box P-7, LLNL; Johnson to Gibson, January 28, 1965.
21. W. J. Slazak, memorandum for the record, January 21, 1965, "Progress of the Tennessee-Tombigbee Study," NCG 65–1107, Plowshare Box P-7, LLNL.
22. Johnson to Gibson, January 28, 1965.
23. Reeves to Slazak, June 10, 1965, "Nuclear Excavation Feasibility Study, Tennessee-Tombigbee Waterway," PPE: BB-824, U.S. AEC, Las Vegas, Nevada, Plowshare Box P-7, LLNL.
24. Slazak to Reeves, July 16, 1965, Plowshare Box P-7, LLNL.
25. Ibid.
26. Reeves to Slazak, September 13, 1965, Plowshare Box P-7, LLNL.
27. Panero to Johnson, March 9, 1965, Plowshare Box P-30, LLNL.
28. Author's notes from Plowshare Box P-23, LLNL.
29. Smith to Nordyke, August 20, 1963, Plowshare Box P-22, LLNL; Nordyke to Higgins, "Proposal for Niger-Upper Volta River Diversion Canal," August 15, 1963, Plowshare Box P-22, LLNL.
30. Harrison to Teller, July 21, 1967, Plowshare Box P-23, LLNL. No evidence indicates whether or not Teller made this visit, but it was at least seriously considered: Teller arranged to take time out from a day at the Pentagon to discuss the matter with Livermore's Gary Higgins. Gen [*sic*] to Higgins, July 24, 1967, Plowshare Box P-23, LLNL.
31. Edward Teller (with Allen Brown), *The Legacy of Hiroshima* (Garden City, N.Y.: Doubleday, 1962), 90.
32. "Chronology of U.S.-Australian Plowshare," March 5, 1969, Plowshare Box P-22, LLNL.
33. Leahy to Johnson, January 4, 1964, Plowshare Box P-27, LLNL.
34. Marshall to Palmer, May 17, 1965, Plowshare Box P-22, LLNL. Lloyd Marshall was a Perth journalist. Palmer worked for Holmes and Narver, the AEC contract engineering firm involved in several Plowshare studies.

35. "First Civilian A-Blast May Be in County," *San Bernadino Sun*, June 26, 1963; "A-Blasts May Help Build Desert Roads," *San Bernadino Telegram*, June 26, 1963; Southwick to MacCaulty, "Stories on Amboy, California, Plowshare Study," July 2, 1963, Plowshare Box P-49, LLNL.
36. "First Civilian A-Blast."
37. Ibid.
38. "Minutes of Meeting—Project Carryall," August 22, 1963, AT&SF Office, Los Angeles, Calif., Plowshare Box P-49, LLNL; U.S. AEC, Sante Fe Railway Company, California Division of Highways, "Project Carryall," LIV-38, December 1963, LLNL; Lawrence Radiation Laboratory, "Preliminary Plan for Project Carryall," UCID 4819, October 13, 1964, DOE doc. NV0177266, NTA.
39. U.S. AEC, Sante Fe Railway Company, California Division of Highways, "Project Carryall."
40. Carryall would thus fulfill two key objectives of Livermore's (1962) five-year plan (see table 4.1), an experiment in "multiple operations" and a demonstration project.
41. "External Exposure, Off-Site," n.d. Handwritten notes (no author named). Plowshare Box P-49, LLNL.
42. "Internal Exposure—Off Site—I 131," rough draft, August 27, 1963, WD:KH:cy, Plowshare Box P-49, LLNL.
43. U.S. AEC, Sante Fe Railway Company, California Division of Highways, "Project Carryall."
44. Ibid.
45. "Summary of the Preliminary Carryall Feasibility Study," included in AEC press release, December 23, 1963, DOE doc. NV0144584, NTA.
46. Reeves to Distribution, "Review of Project Carryall Feasibility Study," January 30, 1964, DOE doc. NV0037967, NTA.
47. M. L. Merritt, "Carryall Safety Review," March 11, 1964, RS 5412/320. In Merritt to Miller, March 11, 1964, DOE doc. NV0126735, NTA. Merritt was also skeptical regarding whether Carryall could be conducted without violating the LTBT.
48. Allen to Roehlk, "Review—Carryall Feasibility Study," May 14, 1964; Coogan to Roehlk, "Project Carryall Feasibility Study," May 14, 1964. Both DOE doc. NV0037968, NTA.
49. Philip to Johnson, April 6, 1964, Plowshare Box P-49, LLNL.
50. Ibid.
51. Lawrence Radiation Laboratory, "Preliminary Plan for Project Carryall."
52. Southwick to distribution, "Phase I—Project Carryall: Public Information Plan," draft copy, December 3, 1964, DOE doc. NV0184646, NTA.
53. Johnson to Kelly, June 9, 1965, DOE doc. NV0075622, NTA.
54. Minutes of the Fourteenth Meeting of the Plowshare Advisory Committee, November 9–10, 1965, Las Vegas, Nev. In S. G. English, "Memo for Chairman Seaborg, et al.: Report on Fourteenth Meeting of the Plowshare Advisory Committee," DOE doc. NV0075671, NTA.
55. Vortman, L. "Nuclear Excavation," March 1969, Sandia Laboratories, Albuquerque, N.M., SC-DC-69-1716, Plowshare Box P-19, LLNL. Rarely have I come across any reasonable explanations for the cancellation of Plowshare projects; commonly, the paper trail simply ends when projects were no longer seriously pursued.
56. John Gofman, "Hazards to Man From Radioactivity," paper presented to the Third Plowshare Symposium, Livermore, 1964, Plowshare Box P-26, LLNL.
57. At that time, primarily in connection with debates over the safety of nuclear power production facilities, Gofman and colleague Arthur Tamplin argued that present permissible dose limits, at 0.17 rads per year, were ten times too high for reasonable assurances of safety. Quotations taken from an interview with Gofman in Carole

Gallagher, *American Ground Zero: The Secret Nuclear War* (New York: Random House, 1993), 300.

58. "Remarks by Richard Hamburger at the Panama Conference on Tropical Biology, Smithsonian Institution, Panama City, Panama," November 10, 1966. AEC press release, November 11, 1966, Plowshare Box P-26, LLNL.
59. U.S. AEC, "Canal Study Commission Comments on Katzenbach's Questions," November 17, 1966, AEC 811/143, DOE doc. NV0075867, NTA; U.S. AEC, "Plowshare Program during 1966," based on annual report to Congress of the U.S. AEC, 1966, Plowshare Box P-28, LLNL.
60. U.S. AEC, "Plowshare Program during 1966."
61. Klement to distribution, "Canal Studies Bioenvironmental Program," April 7, 1966, DOE doc. NV0061970, NTA.
62. Ibid.
63. Battelle Memorial Institute, "Bioenvironmental and Radiological-Safety Feasibility Studies, Atlantic-Pacific Interoceanic Canal," January 13, 1967, DOE doc. NV0016493, NTA; see also R. G. Fuller and W. E. Martin, "Bioenvironmental Safety," n.d. Battelle Memorial Institute, Columbus, Ohio, DOE doc. NV0067955, NTA.
64. Ibid.
65. Battelle Memorial Institute, "Bioenvironmental and Radiological-Safety Feasibility Studies."
66. Ibid.
67. Atlantic-Pacific Interoceanic Canal Study Commission, "Third Annual Report," July 31, 1967, DOE doc. NV0065737, NTA.
68. Ibid.
69. Kenneth Martell, "The Plowshare Nuclear Sea-Level Canal Project," May 1969, 21, National Center for Atmospheric Research, Boulder, Colo. DOE doc. NV0-173161, NTA. Martell's paper, under review at this time at *Environment* (formerly *Nuclear Information*), was made available to the AEC (presumably by Martell himself) in advance of publication.
70. David Perlman, "Canal A-Blast Problems," *San Francisco Chronicle*, September 5, 1968.
71. Ibid.
72. Atlantic-Pacific Interoceanic Canal Commission, "Third Annual Report." Dry seasons in the winter, relatively speaking, made that the preferred blast time.
73. Perlman, "Canal A-Blast Problems." Martin's own work, which incorporated "standard man" into food web studies, drew few conclusions aside from evaluations of the models themselves. See S. G. Bloom, A. A. Levin, W. E. Martin, and G. E. Raines, "Mathematical Methods for Evaluating the Transport and Accumulation of Radionuclides," Bioenvironmental and Radiological-Safety Feasibility Studies, Atlantic-Pacific Interoceanic Canal, April 1, 1970, Battelle Memorial Institute, Columbus, DOE doc. NV0083126, NTA.
74. Deputy Defense Secretary Cyrus Vance in 1966, cited in Trevor Findlay, *Nuclear Dynamite: The Peaceful Nuclear Explosions Fiasco* (Sydney: Brassey's Australia, 1990), 72.
75. These cratering tests will be discussed more in chapter 5.
76. Thomas O'Toole, "Buggy, Cabriolet A-Tests Prove Canal Feasibility," *Washington Post*, May 14, 1968.
77. Atlantic-Pacific Interoceanic Canal Commission, "Fourth Annual Report."
78. Anderson to Seaborg, August 23, 1968. Included in Martell, "The Plowshare Nuclear Sea-Level Canal Project."
79. Martell, "The Plowshare Nuclear Sea-Level Canal Project." The AEC's largest test explosion ever was (and is still) the 15-megaton Bravo test at Enewetak in 1954.

Canal plans called for three 50-megaton explosives, among others, to excavate a cut from 1,100 feet above sea level along the continental divide in Panama.

80. Seaborg to Anderson, August 30, 1968. In Martell, "The Plowshare Nuclear Sea-Level Canal Project."
81. John Kelly, "Statement on Plowshare Program," U.S. AEC, Fiscal Year 1968 Authorization Hearings for the Joint Committee on Atomic Energy, March 1967, Plowshare Box P-28, LLNL.
82. AEC press release, December 10, 1968, Plowshare Box P-29, LLNL.
83. Quoted in United Press, "Report of Fallout Peril in Utah," *San Francisco Chronicle*, December 12, 1968.
84. United Press, "Report of Fallout"; "A-Blasts Endanger West: Prof," *Las Vegas Review Journal*, December 12, 1968.
85. U.S. AEC press release, December 13, 1968, DOE doc. NV0016559, NTA.
86. "AEC Critic Offered Aid by Hughes," *Oakland Tribune*, December 15, 1968. Pendleton declined.
87. Wright to files, "Telecon with Utah Resident RE Schooner Fallout and Benham," AEC Office of Public Affairs, December 16, 1968, DOE doc. NV0016684, NTA.
88. Seaborg is paraphrased here by Frank Toby, who is, in turn, paraphrased in McCraw, Memo to files, "Annual Report to Congress," Division of Operational Safety, December 30, 1968, DOE doc. NV0106450, NTA.
89. Peter Grose, "Radiation Rise in Canada after U.S. Test Is Studied," *New York Times*, January 10, 1969; Times-Post Service, "The Leaky A-Test," *San Francisco Chronicle*, January 9, 1969; Werth to May, January 9, 1969, Plowshare Box P-5, LLNL; U.S. Department of State telegram, "Pravda on Nuclear Fallout in Canada from Ploughshare Tests," January 14, 1969, Plowshare Box P-5, LLNL.
90. What that data eventually showed was far from a hundredfold decrease from Sedan. Sedan had released 1,460 equivalent tons of fission, and Schooner, less than 35 percent the size of Sedan, produced 370 tons (or about 25 percent as much fallout as Sedan), thus producing (proportionally) some 75 percent as much fallout as the Sedan bomb. Howard Tewes, "Results of the Schooner Excavation Experiment," UCRL-71832, January 9, 1970, Lawrence Radiation Laboratory, Plowshare Box P-5, LLNL.
91. Australian Atomic Energy Commission, *Seventeenth Annual Report, June 30, 1969* (Lucas Heights, NSW, Australia: Australian Nuclear Science and Technology Organisation [ANSTO]).
92. Press statement by D. E. Fairbairn, Australian minister for national development, January 23, 1969, Plowshare Box P-23, LLNL.
93. Ibid. To this point, in addition to its unsuccessful try at power production with Gnome in 1961, the only other Plowshare "application" to be executed was in natural gas stimulation. A joint AEC–El Paso Natural Gas Company test was conducted (underground) in New Mexico in December 1967. Although the AEC claimed success, the gas that the explosion had propelled to the surface was in fact much more contaminated with radioactive tritium than expected. On Plowshare's natural gas stimulation program, see F. Kreith and C. Wrenn, *The Nuclear Impact* (Boulder, Colo.: Westview Press, 1976); Findlay, *Nuclear Dynamite*, 181–203.
94. "Radiation No Longer a Problem," *The Australian*, February 13, 1969; David Fairbairn, "Atom Port Plan a First Step," *Sydney Morning Herald*, February 1, 1969; "W.A. Backs Nuclear Excavation," *Sydney Herald*, February 1, 1969; Leonard Bickel, "One Big Bang Will Make New Harbor," *The Australian*, February 5, 1969.
95. Australian Atomic Energy Commission, *Fifteenth Annual Report, June 30, 1967* (Lucas Heights, Australia: ANSTO).

96. Teller to Timbs, March 29, 1967, Plowshare Box P-22, LLNL; Hancock to Teller, March 16, 1967, Plowshare Box P-22, LLNL; Nordyke to Werth, "Discussions with Mr. Lang Hancock, August 18, 1967," August 21, 1967, Plowshare Box P-22, LLNL.
97. Nordyke to Werth, "Discussions with Mr. Lang Hancock."
98. Nordyke to Kelly, March 26, 1969, Plowshare Box P-22, LLNL. According to Nordyke, the procedures for proposing the project had been laid out on a previous Plowshare trip to Australia (by Werth and Kelly, as requested by Teller in March, 1967): "the Australian government officials made it very clear that any company that wanted to proceed with a project should work though the Western Australian government who would in turn notify the Commonwealth government."
99. Nordyke, "Discussion of the Need for a Harbor Demonstration."
100. Ibid.
101. Hosmer to Johnson, December 27, 1969, Plowshare Box P-28, LLNL. For Hosmer, a Republican JCAE member from California, the main problem with Plowshare was perennially that the program was not moving fast enough.
102. Milo Nordyke, "Concept for a Demonstration Harbor at Cape Keraudren, Western Australia," UCRL-50598, February 13, 1969, Plowshare Box P-23, LLNL.
103. "Summary of Harbor Management Meeting, February 14–15, 1969," Plowshare Box P-23, LLNL; "Australian Harbor Feasibility Study: Summary of Understanding," February 18, 1969, Plowshare Box P-23, LLNL; Director's Office, LRL, press release February 18, 1969, Plowshare Box P-23, LLNL.
104. "Australian Harbor Feasibility Study"; "Summary of Harbor Management Meeting."
105. Rodean to Werth, Higgins, and Woodruff, "Australian Harbor Feasibility Study—II: LRL Participation, Tasks, and Foreign Travel," February 25, 1969, UOPKA 69–20, Plowshare Box P-22, LLNL. The nature of the relationship between Sentinel and National Bulk Carriers with "Premier" remains unclear; this is the only document that I found in which Premier's name appears.
106. Seaborg told the Senate Foreign Relations Committee, for instance, that although the Cape Keraudren blast was sure to send radioactive debris beyond Australia's borders, an *understanding* (short of an amendment) could be achieved to make it permissible by the PTBT. Findlay, *Nuclear Dynamite*.
107. Maurice Timbs, "Cape Keraudren Project," notes to file, n.d., Plowshare Box P-22, LLNL.
108. Marvin Kalkstein, "Nuclear Explosives to Create a Harbor" (letter to the editor), *New York Times*, February 23, 1969.
109. Findlay, *Nuclear Dynamite*, 152–153.
110. U.S. AEC and AAEC press release, March 29, 1969, Plowshare Box P-23, LLNL.
111. Roger Allebone, "Nuclear Policy under Fire," *Honi Soit*, April 1, 1969; *Sydney Morning Herald*, "Appeal for Backing on Atom Port Criticism," April 2, 1969.
112. Virginia Brodine, "Unsnug Harbor," *Environment* 11 (May 1969): 34–36.
113. Kelly to Seaborg, Ramey, Tape, Johnson, and Costagliola, "Future of Australian Plowshare Study," April 7, 1969, Plowshare Box P-23, LLNL.
114. Rodean to distribution, "NTS Tour with Australian Newspaper Correspondents Leonard Bickel and Lloyd Marshall," July 24, 1969, Plowshare Box P-23, LLNL. For the consummate booster Marshall, the push for nuclear projects in Western Australia persisted as late as May 1969, with the publication of his article "Nuclear Harbour—Now," which argued for Cape Lambert in Western Australia as a new site for geographical engineering (*Independent Look*, May 4, 1969). As Marshall explained it to Livermore's Gary Higgins, "You cannot forcefeed the multitude with bangs in their ears. Gently, gently catchee monkey." Marshall to Higgins, August 16, 1969, Plowshare Box P-23, LLNL.

6. Epitaph

1. "Harbor on North Slope Dug by Nuclear Blast Is Mulled, Teller Says," *Wall Street Journal,* September 25, 1969.
2. From Peter Coates's environmental history, *The Trans-Alaska Pipeline Controversy* (Bethlehem, Penn.: Lehigh University Press, 1991).
3. The idea was pitched in a series of meetings with oil executives. Nordyke to Werth, March 17, 1969, Plowshare Box P-22, LLNL; Nordyke to Werth, April 21, 1969, Plowshare Box P-22, LLNL.
4. Milo Nordyke, "Discussion of the Need for a Harbor Demonstration," UOPKC 68–106, November 18, 1968, Plowshare Box P-22, LLNL.
5. Brigadier General R. H. Groves, "Status of the Interoceanic Canal Study," Corps of Engineers, Department of the Army, January 14, 1970, Engineering Agent for the Atlantic-Pacific Interoceanic Canal Study Commission, DOE doc. NV0173118, NTA.
6. Ibid.
7. Ibid.
8. The 44-million-dollar request was reported in "Keeping Plowshare Alive," *Nature* 225 (March 14, 1970): 992–993. The Division of Peaceful Nuclear Explosives ultimately requested $36.1 million for Plowshare, and the AEC cut this amount to $22.9 million in its budget for FY 1971. Findlay, *Nuclear Dynamite.*
9. "Statement of Dr. Michael M. May, Director, LRL-Livermore before the Joint Committee on Atomic Energy on March 5, 1970," Plowshare Box P-28, LLNL.
10. In the final analysis, the conversion of the existing canal to a sea-level canal by conventional excavation techniques was *less* expensive than nuclear excavation plans by some $200 million when safety, evaluation, and evacuation plans were included.
11. The stimulated gas was contaminated by radioactive tritium in both Colorado tests. In fact, due in part to a leak in the pipe used to propel tritiated water into a disposal well, U.S. Environmental Protection Agency gas samples recovered from the Rio Blanco project contained one thousand times the normal background radiation levels. Neither test demonstrated the commercial viability of the Plowshare gas stimulation technique—particularly since consumers, such as the city of Aspen, Colorado, proved reluctant to buy the tritium-laced gas despite AEC assurances of safety. F. Kreith and C. Wrenn, *The Nuclear Impact* (Boulder, Colo.: Westview Press, 1976). Plowshare remained on the books at much reduced financial levels until 1977, when the U.S. Department of Energy was created.
12. Herbert Marcuse, *One-Dimensional Man* (Boston: Beacon Press, 1964), 247–248. Internal quotation cited from *New York Times,* November 11, 1960.
13. Interview with Herbert York, August 4, 2004.
14. N. J. Wilimovsky, ed., J. N. Wolfe, asst. ed., *Environment of the Cape Thompson Region, Alaska* (PNE-481, Division of Technical Information, United States Atomic Energy Commission, 1966); Wolfe to Johnson, June 16, 1966, and Johnson to Wolfe, July 1, 1966, both Gerald Johnson Papers, MSS. 206, Box 6, Folder 48, Mandeville Special Collections Library, University of California, San Diego.
15. Gerald Johnson, "Plowshare at the Crossroads," in *Alamogordo plus Twenty-five Years,* ed. R. S. Lewis, and J. Wilson (New York: Viking Press and Educational Foundation for Nuclear Science, 1970), 196–210. Quotation from p. 201. Reprinted from *Bulletin of the Atomic Scientists,* June 1970.
16. Ibid.
17. See, for example, Ulrich Beck, *Ecological Politics in the Age of Risk,* trans. Amos Weisz (Cambridge: Polity, 1995); Andrew Feenberg, *Questioning Technology* (Lon-

don: Routledge, 1999); Bruno Latour, "To Modernize or Ecologise? That Is the Question" in *Remaking Reality: Nature at the Millennium*, ed. B. Braun and N. Castree, trans. C. Cussins (London: Routledge, 1998), 221–242; Steven Yearley, "The Environmental Challenge to Science Studies," in *Handbook of Science and Technology Studies*, ed. S. Jasanoff, T. Pinch, G. E. Markle, J. C. Peterson (Beverly Hills, Calif.: Sage, 1994), 457–479.

18. Ulrich Beck, *World Risk Society* (Cambridge: Polity, 1999), 58.
19. Statement by Dr. Willard Libby, Atomic Energy commissioner, "Plowshare Notes from Geneva Press and Radio Conference, September 11, 1958," Plowshare Box P-29, LLNL.
20. Sometimes monsters may rise from the grave. A photograph of the Sedan explosion graced the November 2003 cover of *Physics Today*, but it was not in an effort to revive the "peaceful nuclear explosion." Rather, under the heading "A need for new nukes?" the article speaks to the resiliencies of experimental cratering knowledge, and the laboratories that produced it, as it may apply to American depleted uranium (DU) and "mini-nuclear" bunker-buster bombs that a new generation of weapons scientists has developed to match a new geopolitical context. Robert W. Nelson, "Nuclear Bunker Busters, Mini-Nukes, and the US Nuclear Stockpile," *Physics Today* 56 (11): 32–37.

Index

About the Author

Scott Kirsch is an assistant professor of geography at the University of North Carolina at Chapel Hill.